Cooking Cosmos
Unraveling the Mysteries of the Universe

Cooking Cosmos

Unraveling the Mysteries of the Universe

Asis Kumar Chaudhuri

Variable Energy Cyclotron Center, India

World Scientific

NEW JERSEY · LONDON · SINGAPORE · BEIJING · SHANGHAI · HONG KONG · TAIPEI · CHENNAI · TOKYO

Published by

World Scientific Publishing Co. Pte. Ltd.

5 Toh Tuck Link, Singapore 596224

USA office: 27 Warren Street, Suite 401-402, Hackensack, NJ 07601

UK office: 57 Shelton Street, Covent Garden, London WC2H 9HE

Library of Congress Cataloging-in-Publication Data

Names: Chaudhuri, Asis Kumar, author.

Title: Cooking cosmos : unraveling the mysteries of the Universe / Asis Kumar Chaudhuri
  (Variable Energy Cyclotron Center, India).

Description: Hackensack, NJ : World Scientific Publishing Co. Pte. Ltd.,
  [2017] | Includes index.

Identifiers: LCCN 2016029040| ISBN 9789813145764 (hardcover ; alk. paper) |
  ISBN 9813145765 (hardcover ; alk. paper) | ISBN 9789813145771 (pbk ; alk. paper) |
  ISBN 9813145773 (pbk ; alk. paper)

Subjects: LCSH: Cosmology.

Classification: LCC QB981 .C5374 2017 | DDC 523.1--dc23

LC record available at https://lccn.loc.gov/2016029040

British Library Cataloguing-in-Publication Data

A catalogue record for this book is available from the British Library.

Desk Editor: Ng Kah Fee

Typeset by Stallion Press
Email: enquiries@stallionpress.com

Dedication

In memory of my father Late Durga Mohan Chaudhuri and mother Late Tarubala Chaudhuri, who instilled the spirit of scientific enquiry in my mind.

Contents

Preface

George Orwell once alluded to four motive forces that compel an author to write. They are: (i) Sheer egoism, the desire to seem clever, to be talked about, to be remembered after death, etc.; (ii) Aesthetic enthusiasm, the desire to share author's perception of beauty — the beauty of rightly arranged words, their rhythm — the desire to share an experience which the author feels valuable etc.; (iii) Historical impulse, the desire to see things as they are, to find out true facts and store them up for the use of posterity; (iv) Political purpose, using the word "political" in the widest possible sense. It is the desire to push the world in a direction which the author thinks is the best, desire to alter other people's idea of the kind of society that they should strive after.

The fourth motive force, political purpose possibly best describes the motive force for the book you are holding, *Cooking Cosmos: Unraveling the Mysteries of the Universe*. It has a purpose: to rekindle men's interest in science. I am a practitioner of science and believe that science — the supreme embodiment of reason — holds the key to human freedom, freedom from poverty, freedom from hunger, freedom from slavery, freedom from religious slavery. Yet, in general, a common man is indifferent to science. His attitude to science is best described by Sir Oliver Lodge's remark, "I (the common man) won't cross the street to learn about it (science)." Over the years, the attitude has not changed much, and even if it has changed, it has changed for the worse. A common man is now more apathetic to science than he was before. What is even more worrying is that throughout the world, religious fundamentalism is taking a grip on the society. The objective of religious fundamentalists is clear: to recover and publicly institutionalize aspects of the medieval past that modern world has rejected.

Possibly, the only way to counter the spread of religious fundamentalism is to increase awareness to science among the world populace, to inculcate scientific spirit among the world's citizens. Religion and science are coeval, the day a man first conceptualized "God," in a sense he became a scientist. "God" is the ancient man's answer to the ill-understood nature. Answer to the questions "How we came to be?" "Why we are here?" etc., the very answers science strives for, but in a different form, and in a different manner. One story may not be inappropriate here. French mathematician and scientist, Pierre-Simon Laplace, on completion of his monumental work, the *Celestial Mechanics*, presented a copy to the Emperor, Napoleon Bonaparte. Bonaparte asked him, "*I understand you did not mention God even once?*" to which Laplace replied, "*Sire, I had no need of that hypothesis.*" Bonaparte in a conversation with Adrien-Marie Legendre, another French mathematician and scientist, told him of Laplace's answer. Reportedly, Legendre said, "*Why? It is not a bad hypothesis.*" Indeed, God, as a hypothesis is not bad. It has served and is still serving human societies. To the ancient men, it provided the binding force to form the clans, groups which ultimately evolved to modern societies. It gave solace to a man in despair. However, when God was institutionalized in the form of religion, it started to conflict with science. By their innate nature, institutionalized religion and science are contradictory. Institutionalized religion is averse to change. To a believer God is the creator and sustainer of the world. Then how religion, which is a divine revelation and the embodies the creator's concept of an ideal World, can change? Does not the omnipotent God has all the knowledge? Science, on the other hand, is progressive; it is not averse to change. Indeed, given irrefutable evidence, scientific ideas on every subject are liable to change. Thus, when found irrefutable evidence (Jupiter's moon), Galileo did not hesitate to abandon the concept of stationary Earth and embraced the Copernican model of Earth rotating around the stationary Sun. Church, on the other hand, could not do so. It rejected the irrefutable evidence and blindly clung to its scripture which said that Earth is at rest. In the 17th century, it appeared that religion and science had marked their sphere of influence. Religion restricting itself to human moral and values and science to the natural world and facts. The humanity prospered in leaps and bounds. Men's way of life drastically changed. What was a sin in the medieval age is no longer a sin today. Institutionalized religions frowned upon the modern way of life and in the twentieth century, throughout the world, groups emerged with the objective to recover and publicly institutionalize aspects of the past that modern life has obscured. Undoubtedly, there are various political and economic reasons for the growth of religious fundamentalism, which I will not dwell upon. However, the goal of religious fundamentalists is clear: abandon the modern way of life and go back to the medieval age, where half of the world's populace, the women will not have an independent existence. They will not have any right, rather they will be the property of the other half, men.

The evil of religious fundamentalism cannot be countered only by economic development or political good will. True scientific spirit needs to be inculcated among the world's populace. Efforts must be made to rekindle men's interest in science. Unlike the religion, which is based on faith, science is based on reason and evidence. It teaches you to ask questions. It teaches you to use your reasoning power to distinguish between truths and lies. A scientifically oriented mind will see through the falsity and lies of fundamentalist propagandas, the evil of religious fundamentalism. As a responsible member of the modern society, with certain knowledge of science, I felt an urge to write a book as my contribution in the process of rekindling men's interest in science. What better way to achieve that than to discuss about the Cosmos, our ordered Universe? Every man, at one time or another in his life must have felt an urge to know more about the Universe, "How it came to be?" "What will happen to it?" This book intends to satiate this urge of mankind. The book will take the reader through the intellectual journey of mankind to unravel the mysteries of the Cosmos. Beginning with Aristotle's Earth-centered Universe, the book takes the reader step by step to the Copernican Sun-centered Universe, to Hubble's expanding Universe, to the Big Bang, to the currently accepted accelerating Universe. In the process, the book explores the origin of space-time, black hole, black hole radiation, dark matter, dark energy, quantum gravity, string theory, etc., all in terms comprehensible to general audiences. The book will also dwell on the question, "What happened before the Big Bang?" It is my belief that to the reader, the book will present an excellent introduction to the way modern science works. I sincerely hope that this book will serve its purpose to increase scientific awareness among the world populace, inculcate scientific spirit among us and help in defeating religious fundamentalism.

In the course of writing this book, I have received help and encouragement from various quarters. Foremost, I would like to thank my brothers, Barun Chaudhuri and Gurupada Chaudhuri, and in-laws, Pijush Kanti Majumder and Manjulika Majumder, for constantly encouraging me in my endeavor. Gurupada patiently read through the book and made various comments and suggestions, which greatly improved the manuscript. I would also like to thank my colleagues and friends: Dr. Saila Bhattacharya, Dr. Santanu Pal and Dr. Y. P. Viyogi. They went through certain parts of the book and suggested improvement. Lastly, the two years of labor that went into the book would not have been possible without the active help and support of my wife Suparna and my son Turja. No word can express my indebtedness to them.

Chapter 1

Introduction

> The first human who hurled an insult instead of a stone was the founder of civilization.
>
> Sigmund Freud

1.1 At the beginning

Far away from the city, unencumbered by pollution and artificial lights, if you look up on a clear night, you will be dazzled by a spectacular sight of diamond studded black sky. Overwhelmed by the sight, you cannot resist from asking: "Who made this?" Indeed, since our first awakening, *Homo sapiens* or humans have been asking: Who made this Universe? Where did it come from? How and why did it begin? Will it come to an end, and if so, how? Throughout the ages, men have been trying to answer these questions.

At the beginning when humans first appeared on the Earth, they observed but did not understand why, with precise regularity, the glorious Sun rose in the east in the morning and blazed its way through the sky to set in the west. They did not understand why, as the Sun set, dark night pervaded his habitat and the softly glowing Moon appeared along with thousands of pinpricks of light. They could only look at the sky and be marveled by its beauty. The vast

1

continent, huge mountains, limitless sea reminded them of their own inadequacy, insufficiency. Rain, storm, flood, thunder, disease threatened them directly. Very early in the history of mankind, ignorance of nature and fear of the unknown made men religious, they conceived of the omnipotent God. God created and controlled the nature, He also controls their life. When God is pleased with men, they are rewarded with abundant sustenance, easy hunting, and gathering; displeased they are punished with rain, flood, thunder and diseases. Initially, in primitive societies, there were many Gods (now still in certain religions). Thus, there were Gods for heavenly bodies like the Sun and the Moon, for fire, rain, thunder etc. Religious men tried to answer the questions asked at the beginning. In their answers, God played a central role. God is the sole creator and sustainer of the Universe. Thus, *Book of Genesis*[1] begins with,

"In the beginning God created the heavens and the earth."

and goes on describing how in six consecutive days God created the world and all its content and rested on the seventh day. In Biblical thought, God created the world *ex nihilo*, meaning out of nothing. One of the oldest religion Hinduism, however, questions the *ex nihilo* creation of the world. Indeed, many scholars believe that the Hindu idea of creation is the most modern and comes closest to the currently accepted cosmological theories. To quote from Carl Sagan's *Cosmos*,

"The Hindu religion is the only one of the world's great faiths dedicated to the idea that the Cosmos itself undergoes an immense, indeed an infinite, number of deaths and rebirths.

It is the only religion in which the time scales correspond to those of modern scientific cosmology. Its cycles run from our

---

[1] *Book of Genesis* is the first book of the Hebrew Bible or the Christian Old Testament.

ordinary day and night to a day and night of Brahma, 8.64 billion years long. Longer than the age of the Earth or the Sun and about half the time since the Big Bang."

The Hindu religious book *Rig Veda* is a collection of hymns from around 1500 to 800 BCE. The hymns are in praise of various Hindu Gods, Agni — God of fire; Varuna — God of air; Indra — God of rain and thunder; and various other Gods. One of the hymns: 10-129, has fascinated intelligentsia around the world. It presents an insoluble paradox: How can the Universe have sprung into existence out of nothing? How can there be a beginning, before which there was nothing? To quote from *Rig Veda* (translations by Ralph T H Griffith),

"Who verily knows and who can here declare it, whence it was born and whence comes this creation? The Gods are later than this world's production. Who knows then whence it first came into being? He, the first origin of this creation, whether he formed it all or did not form it. Whose eye controls this world in highest heaven, he verily knows it, or perhaps he knows not."

Over time, with intellect and experience men could understand more and more of the natural occurrences and slowly we became scientific. As we became more and more scientific, God gradually became less and less relevant in many aspects of our lives. However, it will be wrong to believe that men have become religion-free. Religion appears to be incredibly resilient, resisting the most determined attempts by powerful states like Russia, China etc. to repress and extinguish it. Some social scientists believe that human beings are religious by nature.

Thousands of years of continual interaction with nature brought mankind to the present stage when we have some inkling about its working. Incomplete may be the knowledge yet it is sufficiently advanced to unravel some deep mysteries of the nature. We have made some astonishing and unexpected discoveries about the Cosmos and our place within it. We now know that the Earth, our habitat is only one of the planets of the Sun. The Sun itself is a star among billions of stars in the Universe. We know that our solar system came into existence some 4,500,000,000 (4.5 billion) years ago and it is but only a tiny component of a galaxy[2] called Milky Way. Our Universe contains some 100 billion galaxies. We know that the Universe itself came into existence with the *Big Bang* some 13.7 billion years ago and even now galaxies are receding from each other with ever increasing speed. An ordinary man wants to know how such understandings came about. To satiate one's yearning for knowledge, we will take an intellectual journey in the human endeavor in unraveling the mysteries of the Cosmos.

1.2 Early history of mankind

4.5 billion years ago when Earth was formed by a quirk of nature, it was devoid of any life. Indeed, Earth at birth was completely unlike the Earth we now know. It was hot, liquid and suffered continual bombardment from meteorites.[3] Over the years, the bombardments subsided, the Earth cooled, surface solidified and approximately 3.8 billion years ago, life in some primitive form e.g. single cell amoeba or bacteria, appeared on the Earth. It took a billion year for the single cell life to evolve into multi-cell life. The exact mechanism by which this evolution proceeded is not properly understood. Atmospheric and geological changes on Earth must have forced the single cell organisms to evolve into multi-cell organisms through the process of natural selection.[4] The evolution to multicellularity required alignment

---

[2]A galaxy is a gravitationally bound system of stars, stellar remnants, interstellar gas, dust, dark matter etc. The Universe contains billions of galaxies.

[3]A meteorite is a fairly small natural object from interplanetary space that survives its passage through Earth's atmosphere and lands on the surface.

[4]Charles Robert Darwin gave the theory of biological evolution where all species of organisms arise and develop through the natural selection of small, inherited variations that increase the individuals' ability to compete, survive, and reproduce.

of fitness; the cells must adhere to, communicate with, and cooperate with each other. It also required export of fitness that the cells work together for the common goal of reproduction, producing more multi-cell units in the image of the parent. The kind of life we are familiar with appeared much later. For example, fish appeared on Earth approximately 530 million years ago (Ma). Mammals didn't evolve until 200 Ma. Our own species, humans or *Homo sapiens*, appeared on Earth only 200,000 years ago. Humans are descendants of the Great Ape family — Hominid, who were characterized by bipedalism, i.e. the ability to walk on rear limbs or feet. In the "Out of Africa" theory of human evolution, around two million years ago, *Homo habilis* of genus *Homo* first appeared in the African region of the Earth. In Latin, *Homo* means "man" and *habilis* means "handy." They were the close relation of the modern human though retained some ape-like features e.g. long arms and protruding jaw. They were the first makers of stone tools, hence the name "handy man." Over the years, *Homo habilis* evolved into *Homo erectus*, who used to walk upright. Now extinct, *Homo erectus* inhabited Earth during 1.9–0.07 Ma. They migrated across Africa to Eurasian region, going as far as India, China, and Indonesia. Over the years, around 0.36–0.04 Ma, *Homo erectus* branched into two species: (i) *Homo neanderthalensis* or Neanderthals, now extinct; and (ii) *Homo heidelbergensis*. Around 0.2 million years ago, *Homo heidelbergensis* evolved to *Homo sapiens* — human beings or modern man.

Man is the key element of the world. Among the millions and millions of planets in our Universe, possibly only Earth has the distinction that it gave birth to this beautiful creature, who can think. Archeologists have divided the period of human history on Earth into three eras, prehistoric, proto-historic and historic. The historic era is the period when men have mastered the art of writing. Evidence of writings was obtained from various excavations in modern time (within the last four to five hundred years). The prehistoric era spanned a long time and can be divided into "Paleolithic", "Mesolithic" and "Neolithic" ages. In Greek, "*Lithos*" means stone and the eras are associated with different stages of stone implements or instruments used by the mankind. *"Palaios"* in Greek means old or ancient. In Paleolithic or "old stone" age, mankind was in its most savage form. They had to cohabit the planet with other animals and survived against immense odds. Earth was not a safe place to them and considering the physical structure, it is amazing that mankind survived and even prospered. A man is not strong as an elephant or a rhino, cannot run fast as a dear, he is also devoid of fierce teeth and nails of a lion. The only thing he was superior to his cohabitants was intelligence. He surpassed his physical weakness with superior intelligence and countered and prospered against his formidable opponents. In Paleolithic or Old Stone Age, men lived in caves and in groups (to scare away the wild animals). They used simple stone implements to kill as well as fend off adversaries. They obtained their sustenance from hunting and gathering. It is difficult to say when, but at some stage towards the middle of Paleolithic era, men learned the use of "fire." They were familiar with fire, having witnessed thunder striking trees to set them ablaze. The achievement was to light the fire as and when required and to control it. Harnessing the power of fire advanced men variously. They became less vulnerable to their adversaries; wild animals are always scared of fire. Fire provided a source of warmth, comfort in cold nights. It allowed men to extend their activities during the night times. It also made them healthier, they could then eat cooked meat, not only more palatable but, unknown then, free of unhealthy bacteria e.g. *E. Coli* and salmonella.

The Paleolithic age continued till 10,000 BCE and mankind entered the Mesolithic era. In Greek *"Mesi"* means middle. In the Mesolithic or Middle Stone Age, mankind gradually improved their stone implements to make them more effective and lethal, e.g. sharpening stones by rubbing on hard rocks or launching them with a sling. They also learned to make semi-permanent houses using woods and leaf. They learned to fish and augmented their source of food and possibly started using the boat. The Mesolithic age continued possibly till 5000 BCE. Towards the end, Mesolithic men developed agriculture. Developing agriculture propelled them far ahead of other inhabitants of the planet. It changed their living pattern completely. Before farming or agriculture, men were nomadic in nature, moving from place to place in search of food and herding for their animals. Once they could produce their food, they started to settle in and around fertile river valleys, where water — the most essential ingredient for farming — is assured. All the ancient civilizations were river bank civilizations e.g. Mesopotamian civilization (part of modern Iraq) between Tigris and Euphrates, Indus valley civilization (now in Pakistan) along the river Indus, Egyptian civilization along the Nile river and Huan He civilization along the river Huan He in China.

In the Neolithic or New Stone Age, men advanced very rapidly. They developed pottery to store grains. They domesticated goats, sheep, donkeys, and other animals. While in Paleolithic and Mesolithic ages family structure was not defined, in the Neolithic age, the concept of the family emerged. Men also learned the use of metals. Neolithic era continued until 2000 BCE and evolved into the historic era when mankind learned the art of writing. Proto-historic era refers to the period between prehistory and historic era. In particular, in certain civilizations, e.g. Indus valley civilization, even though advanced substantially, yet did not learn the art of writing. They communicated their knowledge by oral transmission.

Towards the end of the Neolithic age, progress in metallurgy was substantial and men learned to make tools of copper, bronze and even of iron in some areas. Use of metal played an important role in the development of civilization and Danish archeologist Christian Thomsen proposed to divide men's history on Earth into three stages according to the use of metals: (i) the Stone Age (up to 3000 BCE), (ii) the Bronze Age (3000–1000 BCE) and (iii) the Iron Age (1000 BCE to present). Thomsen was the curator of the National Museum of Denmark, Copenhagen, and was in charge of the museum's vast antique collections. He arrived at his nomenclature in the course of classifying and arranging the museum's large collection of Scandinavian antiquities. The word metal originated from the Latin word "*metallum*" meaning quarry or mine, indicating that they have to be mined or found in quarries. In the Stone Age use of metal was unknown. In the Bronze Age, men learned the use of metal and progressively advanced to make bronze using copper and tin. In the Iron Age, men advanced enough in technology to smelt iron.

The first encounter of mankind with metal must have been accidental. They possibly found nuggets of gold and copper (these two metals can occasionally be found free in nature) and were attracted by their color and luster. The earliest use must have been ornamental only. With time, they learned that copper and gold were malleable and unlike stones, could be beaten flat. They quickly learned that copper could be beaten to make sharp-edged weapons. However, free copper was rare. Widespread use of copper was not possible. How men learned that copper could be found by smelting some stone (an ore of copper) is not known. Also not known is how men stumbled upon the secret of bronze. Bronze is an alloy of copper and tin. Possibly, on some day, men made a fire on some bluish stone (an ore of copper) and brownish black stone (an ore of tin) and later found the bright metallic bronze. Soon men started to use bronze to make arms and armors. Men had to wait for thousands of years for the next stage of development, the Iron Age.

Iron is not found in pure form and iron ores are more strongly bonded than copper or tin ore. They require much higher temperature for smelting. For example, copper ore requiring 1000°C, can be smelted in a wooden fire; however, a simple wood fire will not be sufficient to smelt iron ore. Iron ore requires 1500°C and requires charcoal fire with good ventilation. Incidentally, Iron Age did not begin simultaneously across the entire world. Earlier, it was believed that systematic production and usage of iron began in the Anatolian region (now in Turkey) around 1200–1000 BCE. It then gradually diffused to other regions: Indians learned the iron technology around 1000–800 BCE, Chinese much later around 600 BCE. However, recent findings in central India indicate that iron smelting and usage were prevalent in Gangetic plains of India as early as 1800 BCE.

I will close this section with a few words on language. *Ethnologue*, published by SIL International (formerly known as Summer Institute of Linguistics), is widely regarded as the most comprehensive source of information on languages. It contains information on 7102 known living languages though more than 2000 of these languages have fewer than 1000 speakers. Unlike animals, all humans have language for communication. Many scholars believe that communication via speech is uniquely human. It is the singular, and most important, dividing line between humans and animals. Admittedly, many animals can communicate by uttering sounds, however, there is a marked difference between tweet of a bird or neigh of a horse and a man standing before an audience and reciting Rabindranath Thakur's[5]:

"Where the mind is without fear and the head is held high…"

When and how humans learned to communicate via speech? When did they develop language? In most religions man is not credited with the invention of language, rather a divine source provides humans with language. Thus in *Book of Genesis* we find,

"Thus God created the man in his image, in the image of God created he him; he created them male and female."

The first human beings, Adam and Eve were then bestowed with the qualities of God, one of which was the ability to speak. In Hinduism, Sarasvati or Vac Devi is the Goddess of speech. In *Rig Veda*, there are numerous references of Goddess Sarasvati. She is clearly

---

[5] Rabindranath Thakur was a Bengali poet, and the first non-European to win the Noble Prize in Literature in 1913 for "his profoundly sensitive, fresh and beautiful verse, by which, with consummate skill, he has made his poetic thought, expressed in his own English words, a part of the literature of the West."

associated with the Sarasvati River. For example Hymn:2.41.16 of *Rig Veda* allude to,

> "Best Mother, best of Rivers, best of Goddesses, Sarasvati, We are, as 'twere, of no repute and dear Mother, give thou us renown."

Later Vedic literature (the Brahmanas) associated Sarasvati with the Goddess of speech. In Sanskrit "Vac" means speech and she was given the name "*Vac* Devi." Even in modern India, Sarasvati or Vac Devi is worshipped as the Goddess of learning.

The origin of language, how and when human acquired the capability to talk, has been a subject of scholarly discussions for several centuries. However, it remained a mystery so much so that in 1866, Linguistic Society of Paris, a foremost academic linguistic institution at the time, banned papers on the origin of languages. The basic difficulty with studying the origin of language is the scarcity of direct pieces of evidence. Spoken languages don't leave fossils. Fossil skulls do tell us about the shape and size of brains, but not what the brains could do. The origin of language is also a major hurdle to Darwin's evolutionary theory. The animal that comes closest to producing anything that even vaguely resembles human speech is not another primate, but rather a bird. There are several speculative theories on origin of language. They can be broadly categorized into two types: (i) "Continuity theories" build on Darwinian perspective that language has evolved from more primitive forms of animal communication and (ii) "Discontinuity theories" with the opposite approach — that language is a unique trait of human and must have appeared fairly suddenly during the course of human evolution. I will not elaborate on these theories. I will only say that the origin of language is now considered the most difficult problem in science.

1.3 Science in ancient civilizations

Before agriculture, the mere task of living was so great that mankind could not spare any time to ponder over nature. In the Neolithic era, men, well settled with agriculture and husbandry, could spare some of their time to ponder and contemplate over nature. Over the years, with contemplations and experiences science grew. The term "science" originated from the Latin word *Scientia* meaning knowledge. Oxford dictionary defines science as:

> The intellectual and practical activity encompassing the systematic study of the structure and behavior of the physical and natural world through observation and experiment.

The purpose of science is to develop general laws that explain how the world around us works and why things happen the way they do. This is done by systematic observations, experiments, and *logical interpretations.*

The exact progress of science in ancient civilizations is difficult to gauge. Considering the engineering feats they have achieved, e.g. the Stonehenge (the huge stone structures built during 3000–2000 BCE) in Salisbury, England; pyramids in Egypt (built during 2700–1700 BCE), use of calendars (used as early as 3000 BCE), it is reasonable to assume that different cultures in the ancient world had certain knowledge of astronomy and mathematics. Indeed, mathematics and astronomy are the two branches of science that grew very early in the civilization.

The word "mathematics" originated from the Greek word "*màthêma*" meaning knowledge, study and learning. Greek philosopher Aristotle defined mathematics as the science of numbers. Today mathematics is defined as the science of numbers, quantities, shapes, and the relations between them. Men settled with agriculture and husbandry needed to learn mathematics. It was required to count their domesticated animals, to measure the plot of land, to fix taxation, etc. Ancient men instinctively knew the differences between say one cow and two cows. But the use of a symbol to designate "two" of anything required an intellectual leap which took many ages to come about. All the ancient civilizations learned mathematics, but it was first introduced in the Sumer region of Mesopotamian Civilization. Sumer is often referred to as the Cradle of Civilization. Sumerians were the first to use a script (Cuneiform) for writing. They were first to use a wheel. The first plow for agriculture was also invented by the Sumerians. It now appeared that even in 2000 BCE, Sumerian mathematics was quite developed. They used a "sexagesimal" or base 60 numeral system. The base 60 system was adopted possibly for the ease of counting up to 60 using 12 knuckles of four fingers of one hand and five fingers of the other. Also, Sumerian did not have the radix point, the symbol we use to separate the integer part from the fractional part of a number. 60 is a super composite number. It is divisible by 1, 2, 3, 4, 5, 6, 10, 12, 15, 20 and 30. Use of fractions is minimized in sexagesimal system. Sumerians also developed a method to describe very large numbers. They also mastered the basic geometry. Over the years, their knowledge of mathematics was passed on to other ancient civilizations e.g. Egypt, Babylon, India, etc. All these civilizations contributed to the growth of mathematics. For example, while the Sumerians and Babylonians used sexagesimal numeric system, Egyptians and Indians used the decimal or base 10 numeric system, the system which is prevalent today. Possibly, Egyptians invented the base 10 system for ease of counting up to 10 using 10 fingers of the two hands.

The word "astronomy" itself indicates its subject of study. It came from two Greek words, "*Astro*" meaning star and "*Nomia*" meaning law or culture, literally meaning culture of star. Astronomy is the science of study of celestial objects like stars, planets, galaxies, etc. Earliest recorded astronomical observations date back to 1600 BCE. The Babylonians kept a detailed record of planetary positions, times of eclipses, etc. Development of astronomy in ancient civilizations also came from a need. Men were religious. Astronomical bodies, such as the Sun, Moon, and planets, were often seen as Gods and Goddesses. Predicting planetary positions were a way of predicting the will of the Gods. Eclipses and comets[6]

---

[6]A comet is an icy small solar system body that, when passing close to the Sun, heats up and begins to outgas, displaying a visible atmosphere or coma, and sometimes also a tail.

were two of the most important phenomena to ancient people. They were believed to bring devastation and sorrow for men. Furthermore, religious men needed to fix the dates for religious functions, ceremonies. Also, very early, they understood that changes of seasons were related to the motion of Sun. By observing the sky, they would know when to plant a crop and when to reap.

Ancient civilizations also made progress in medicine. Medicine is the science of diagnosis, treatments and prevention of diseases. In ancient times, illness, in general was attributed to witchcraft, adverse astral influence, or the will of the Gods. Ancient men believed in performing various religious rites to cure. However, medicinal values of certain herbs and plants were known to ancient men. Most of the times, religious rites were accompanied with administrations of extracts of plants and herbs. Egyptian medicine was highly advanced as early as 3000 BCE. Homer, best known as the author of *Iliad* and *Odyssey* and who lived around 850 BCE, remarked that "*In Egypt, the men are more skilled in medicine than any of human kind.*" Evidence is there that Egyptians had knowledge of anatomy and performed simple surgeries, bone setting, etc.

However, the use of calendar very early in human history (e.g. from the alignment of the Stone Henge, many believe that it was used as a calendar) speaks voluminously of the intelligence of the ancient men. The word "calendar" came from the Latin word "*Kalendae*" meaning account book. It is the account book of time. Use of calendars then means that very early in the civilization, ancient men had conceived the notion of time. The concept of time is not easy. It is best exemplified by St. Augustine's[7] celebrated quote:

> "What then is time? If no one asks me, I know what it is. If I wish to explain it to him who asks, I do not know."

---

[7] St. Augustine was a Christian theologian and philosopher. His writings greatly influenced Christianity and western philosophy.

Indeed, the question "*what is time?*" cannot be answered in simple terms. In modern science *time* is associated with another difficult concept *space*. The current view is that space and time are not different, but a single entity called space-time. However, in ancient times and even as late as the nineteenth century, these two held to be distinct entities. Space and time are conceptually difficult because they go beyond our sense perception. The human body has five sense organs: eyes, ears, tongue, skin, and nose. We experience the world with these sense organs. However, these sense organs do not experience space or time. We may have some inkling of space — anything that exists is somewhere, in some place. It occupies some space. However, that anything is not space. For example, say a vessel is filled with water. We know that space inside the vessel is occupied with water. Now pour out the water. The space now is filled with air. However, neither water nor air is space. The concept of space requires abstract thinking or intelligence. Space exists in our consciousness, irrespective of material objects. There are several more philosophical issues regarding space. Is it finite or infinite? Is it absolute or exists only in relation with objects? However, we will not dwell on those issues.

The concept of time is similarly abstract. It is intimately related to change. We do not perceive time rather change in time. Change is integral to human experiences: Sun rises in the morning and sets in the evening; the Moon changes its position at regular intervals; plants and animals and humans come into existence, grow, fade and pass away. We remember the changes because we have memory and we remember the past. Though we do not perceive time itself, we can perceive the flow of time. At each instant, the present is turning into past and future is turning into the present. The concept of time allows us to order occurring or events from the past to present to future. Like space, it also exists in our consciousness.

Very early, men knew that some events are cyclic. For example, sunrise in the east is repetitive; every morning Sun rises in the east.

Phases of the moon are also repetitive. From one full moon to another, always a definite time span is elapsed. Similarly, seasons are repetitive. They then devised a way of measuring time by observing heavenly bodies. One sunrise to another or one night to another, they called a day; one full moon to another (approximately 29.5 days) they called a month. They noted approximately every 12 months seasons repeated themselves. They called 12 months a year. At the very beginning ancient men used simple lunar calendar for time keeping, 12 alternate 29 and 30-day months for one year. A lunar calendar year has $12 \times 29.5 = 354$ days. Now we know that Earth rotates about the Sun in 365 days. This gap of 11 days accumulated and over time, each of the twelve months steadily slipped back and the seasons failed to repeat at regular intervals. Occasionally a month had to be added to synchronize with a solar calendar. What was the man's compulsion to divide a month into 4 weeks with 7 days a week is not known. Ancient men were religious. Very early they could identify seven heavenly bodies: Sun, Moon, Mars, Mercury, Jupiter, Venus, and Saturn. They associated a God with each heavenly body. One viable theory is that seven days a week was devised to set aside a day for performing religious rites for each God associated with the heavenly bodies. Indeed, such an origin seems viable when we consider the English and Latin name for the days of the week (listed in Table 1.1, along with the seven heavenly bodies and the associated Gods).

However, true science did not begin until 600 BCE. Science is the human endeavor to find the general laws of nature using systematic observations, experiments and logical interpretations. Although many cultures like the ancient Egyptians and Mesopotamians had collected observations and facts, they had not tried to use those observations and facts to develop explanations for the world around them. Indeed, that didn't happen until the 6th century BCE. While all the ancient civilizations contributed to the growth of science, contribution of Greek's far surpasses that of others.

Table 1.1. Celestial bodies associated with seven days of a week are noted, along with Latin name and Greco-Roman Gods associated with them.

Celestial body	Greek deity	Roman deity	Days of the week	Latin name for the days
Sun	Helios	Sol	Sunday	*dies Solies*
Moon	Selena	Luna	Monday	*dies Lunae*
Mars	Ares	Mars	Tuesday	*dies Martis*
Mercury	Hermes	Mercury	Wednesday	*dies Mercurii*
Jupiter	Zeus	Jupiter	Thursday	*dies Iovis*
Venus	Aphordite	Venus	Friday	*dies Veneris*
Saturn	Cronus	Saturn	Saturday	*dies Saturni*

The first holistic view of the world was taken by the Greek philosophers and modern society is largely indebted to the ancient Greeks. The Greeks introduced the cherished concepts of citizen's rights, democracy and freedom of speech. To the Greeks, modern world owes the Olympic games. The Greeks also contributed enormously to the western art, architecture, mathematics, physics and astronomy. More importantly, the Greeks introduced the notion of scientific thinking and observations. Around 1050 BCE, Greeks migrated to Anatolian peninsula (now in Turkey) and established Ionian city states (the most significant being Ephesus, Miletus and Samos). Over the years with external influences in particular from Egypt and Mesopotamia, the city states prospered, not only in trade and commerce but also in the intellectual arena. Ionian civilization reached its pinnacle around 550–650 BCE and gave birth to a new generation of the philosophers of nature or world's first generation of scientists. These scientists or philosophers were driven by the desire to understand the nature in simplest possible terms. They started the notion of examining the nature free from the effects of religious beliefs and superstitions. To give a few examples,

Heraclitus of Ephesus (ca. 535–475 BCE) insisted on ever changing Universe and famously said, "No man ever steps into the same river twice." Thales of Melitus (ca. 624–546 BCE), took the first holistic view of the world that the origin of everything in nature is the single material substance: water. Everything was made of water and ultimately perished into water. Thales was also the first person to predict a solar eclipse. Greek philosopher Empedocles (ca. 490–430 BCE) however had a different idea. He declined to accept that the world was made of water and was the first to propose four constituents or roots — fire, earth, air, and water — different proportion of which made up everything in the world. Anaximander (ca. 610–546 BCE), also of the Ionian city of Miletus, a disciple of Thales, took the giant leap in mental thinking process when he conceived the "apeiron," the infinite, unlimited or indefinite. He postulated eternal motion, along with the apeiron, as the originating cause of the world. Aristarchus of Samos (ca. 310–230 BCE) presented the first known model that placed the Sun at the center of the known Universe with the Earth revolving around it. They were the early scientists of our world.

Einstein once commented that the most incomprehensible thing about the Universe is that it is comprehensible. However, that comprehension came slowly over thousands of years. In the following pages, I will recount the history of evolution of scientific thought as regard to our Universe. It is an interesting story. I hope that you will enjoy it.

Chapter 2

Classical Cosmology

I know that I am mortal by nature, and ephemeral; but when I trace at my pleasure the windings to and fro of the heavenly bodies I no longer touch the earth with my feet: I stand in the presence of Zeus himself and take my fill of ambrosia.

Ptolemy

2.1 Greeks' view of the world

The ancient Greeks also looked into the night sky and saw those thousands of pinpricks of light. They could see that most of the tiny lights are fixed relative to each other, but a few wander (move) about. They called those wandering lights "planets" (in Greek, planet means wanderer); the rest, fixed tiny lights were called "stars." With their naked eyes, they could see only five planets. The English names by which we now know these five planets — Mercury, Venus, Mars, Jupiter and Saturn — were derived from the names of mythological Roman Gods. In Roman mythology, Mercurius, son of Jupiter, is the patron god for financial gain and moves fast. Planet Mercury was named after him. The name was appropriate, for Mercury, being the innermost planet, orbits the Sun fast, in about 88 days. After the Moon, the brightest natural object in the night sky is Venus. It was appropriately named after the Roman Goddess of love and beauty. Planet Mars, because of its red, bloodlike color, was named after the Roman God of war. The largest and most massive of the planets was named Jupiter after the most important Roman deity Jupiter. Saturn is named after the Roman God of agriculture and harvest.

However, it will be an injustice to the Babylonians if the credit of discovery of planets is given to the Greeks. As early as 2000 BCE, Babylonian used to record celestial movements. They knew of the five planets and called them,

Marduk (Jupiter): the patron God of city of Babylon,
Ishtar (Venus)   : Goddess of love, war, fertility, and sexuality,
Ninurta (Saturn) : God of hunting and war,
Nabu (Mercury) : God of wisdom and writing,
Nergal (Mars)   : God of warfare.

Also, there is no reason to believe that one of the oldest and highly advanced civilizations of the world, the Indus Valley civilization (also called Harappan civilization) did not have any knowledge of the planets. Indus civilization by the river Indus, built around 2600–1900 BCE, had advanced town planning, baked brick houses, elaborate drainage systems, water supply systems, and clusters of large non-residential buildings. Evidence is there that they had advanced metallurgy and produced copper, tin, and bronze. Various sculptures, seals, pottery, gold jewelry, and anatomically detailed figurines in terracotta and bronze speak voluminously of the advanced state of the civilization. Sir John Hubert Marshall, the archeologist responsible for the excavation and discovery of Indus valley civilization, was amazed when he saw the famous Indus bronze statuette of a slender-limbed dancing girl in Mohenjo-daro (see Figure 2.1). He wrote,

"When I first saw them I found it difficult to believe that they were prehistoric; they seemed to completely upset all established ideas about early art, and culture. Modeling such as this was

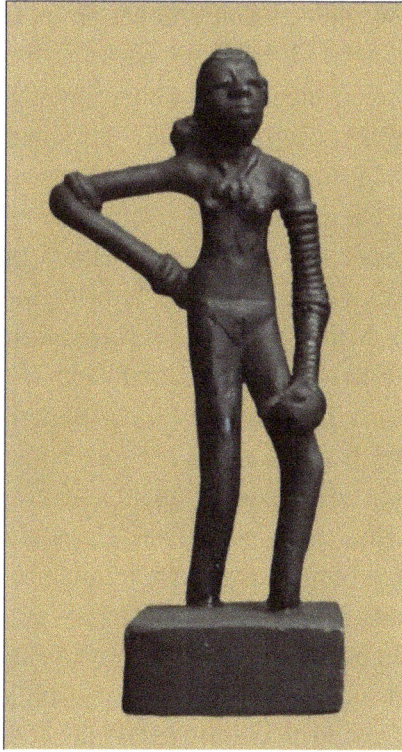

Figure 2.1. The dancing girl of Mohenjo-daro. The image courtesy of Joe Ravi under CC-BY-SA 3.0.

unknown in the ancient world up to the Hellenistic age of Greece, and I thought, therefore, that some mistake must surely have been made; that these figures had found their way into levels some 3000 years older than those to which they properly belonged."

Even though highly civilized, archeologists could not produce any evidence of astronomical knowledge in the civilization. Indeed, historians are perplexed by the Indus valley civilization. How can such a highly civilized society be devoid of any cosmogenic culture? Surely they must have some means of counting time, which will require them to observe the sky. One problem is that the Indus script, comprising around 3500 seals, despite many attempts are still undeciphered.

The Greeks also viewed Sun and Moon as God and Goddess. They associated their God "Helios" with the Sun. People viewed Helios as a mighty charioteer, driving his flaming chariot from east to west across the sky, each day. They associated Moon with the Goddess "Selene." In Greek mythology, she was sister of the Sun God Helios and drove her moon chariot across the heaven. At that time, they had no idea that the Earth is also a planet. The Greeks also saw the patches of dim glowing light arcing across the night sky, they call it the Milky Way. The term is a literal translation of the Latin *via lactea*, which originated from Hellenistic[1] Greek. Greek legend explains the name. God Zeus fathered a son, Heracles, off a mortal woman. Being fond of his son, he wanted him to have devine qualities and surreptitiously let him suckle on his divine wife Hera's milk while she was asleep. When Hera woke up and realized that she was breastfeeding an unknown infant, she pushed him away and the spurting milk became the Milky Way.

While some Greek philosophers believed that Milky Way is a vast collection of stars, the great Greek philosopher Aristotle (384–322 BCE), whose view dominated till the 15th century, believed it to be an earthly phenomenon. Aristotle was a disciple of the famous Greek philosopher Plato (ca. 424–348 BCE) who in turn was one of the disciples of Socrates (ca 469–399 BCE). The trio, Socrates, Plato, and Aristotle greatly influenced the western philosophy and physical science. However, contributions of Aristotle to physical science — pure mathematics and natural science e.g. physics, biology, astronomy, etc. — is far and wide and surpass that of Socrates or Plato. Aristotle was born in 384 BCE at Stagirus, now an extinct Greek colony. His father was a court physician to King of Macedonia. At the age of 17 he went to Athens and enrolled in the world's first center for higher learning, the Academy, which was founded by Plato. At

---

[1] The Greek and Mediterranean history during the period between the death of Alexander the Great at 323 BCE and emergence of Roman Empire at 31 BCE is called the Hellenistic period. Greek culture was at its peak during this period.

the Academy, students were groomed to be a good citizen. Special emphasis was given to mathematics, and reportedly, at the entrance of the Academy, Plato inscribed the following:

"Let no one ignorant of geometry enter."

In Plato's Academy, Aristotle studied for twenty years. He was a gifted scholar but opposed some of Plato's teachings. Upon Plato's death, disappointed that he was not selected as the head of the Academy, he left Athens and spent five years on travelling to Asia Minor (now Turkey). He went back to Macedonia at the invitation of King Philip to teach his 13-year-old son Alexander (later the world conqueror). For five years, he taught Alexander. Both Alexander and his father paid Aristotle high honor, and there are stories that Aristotle was supplied by the Macedonian court, not only with funds but also with thousands of slaves to collect specimens for his studies in natural science. Upon the death of King Philip, Alexander succeeded to the kingship and subsequently went for his world conquest. Aristotle returned to Athens and founded his own academy. In his academy, he taught almost every subjects: logic, physics, astronomy, meteorology, zoology, metaphysics, theology, psychology, politics, economics, ethics, rhetoric, and poetics. A majority of these subjects were not taught before and he was the first to conceive and establish them as a regular discipline. Following the sudden death of Alexander in 323 BCE, the pro-Macedonian government in Athens was overthrown. Fearing persecution, Aristotle fled from Athens to Chalcis of Euboea, the second largest island of Greece. Reportedly, Aristotle said, *"The Athenians must not have another opportunity of sinning against philosophy as they had already done in the person of Socrates."*[2] Within a year, he died at Chalcis.

---

[2] Socrates was accused of corrupting the minds of the youth of Athens and impiety (not believing in the Gods of the state). Athenians sentenced him to death by drinking a mixture containing poison hemlock.

Aristotle was a prolific writer. He extensively wrote on almost every subject: poetics, politics, physics, metaphysics, ethics, rhetoric and more. In early 340–350 BCE, Aristotle wrote two books titled, *Physics* and *On the Heavens*, which were extremely influential, and dominated science until the time of Galileo Galilei. In his book, *On the Heavens*, Aristotle made the first attempt to describe the Universe. He argued that Earth is a round object. The statement, Earth is a round object may seem innocuous today, but it was not always so. The following story from Stephen Hawking's 1988 book *A Brief History of Time* amply demonstrates the situation at early time:

> A well-known scientist (some say it was Bertrand Russell) once gave a public lecture on astronomy. He described how the Earth orbits around the Sun and how the Sun, in turn, orbits around the center of a vast collection of stars called our galaxy. At the end of the lecture, a little old lady at the back of the room got up and said: "What you have told us is rubbish. The world is really a flat plate supported on the back of a giant tortoise." The scientist gave a superior smile before replying, "What is the tortoise standing on?" "You're very clever, young man, very clever," said the old lady. "But it's turtles all the way down!"

Indeed, it is not easy for a man to conceive the true shape of the Earth. In everyday experiences, we perceive Earth only as a flat surface. Early Greek philosopher Pythagoras and also Plato (teacher of Aristotle) taught their students that Earth was round but offered no explanation. The first scientific explanation came from Aristotle. Aristotle understood that lunar eclipses occurred due to blocking of the Sun's rays by the Earth. He observed that the shadow of Earth on Moon during a lunar eclipse was always round, which was only possible if Earth itself was a round object. He wrote:

> "How else would eclipses of the Moon show segments shaped as we see them? As it is, the shapes which the Moon itself each month shows are of every kind straight, gibbous, and concave — but in eclipses the outline is always curved: and, since it is the

interposition of the Earth that makes the eclipse, the form of this line will be caused by the form of the Earth's surface, which is therefore spherical."

There are other arguments as well, e.g. Pole star appears lower in the sky in southern region than in northern region, the mast of an approaching ship is seen first, and then its hull, etc. To speak correctly, Earth is not exactly of spherical shape. Its radius varies from 6356 km at polar region to 6378 km at the equator. The variation is too small and for all practical purposes, spherically shaped earth is a good approximation.

Aristotle, like his teacher Plato, believed in Greek philosopher Empedocles' idea that the entire world is made of four roots or elements: earth (land), air, fire and water. Aristotle endowed them with four forms: hot, dry, wet and cold. Fire is primarily hot and secondarily dry. Air is primarily wet and secondarily hot. Water is primarily cold and secondarily wet. Earth or land is primarily dry and secondarily cold. The classic relation between the elements and forms is shown in Figure 2.2.

Now we do know that Aristotle and the ancient Greeks were wrong in their surmise; earth (or land), water, air and fire are not the

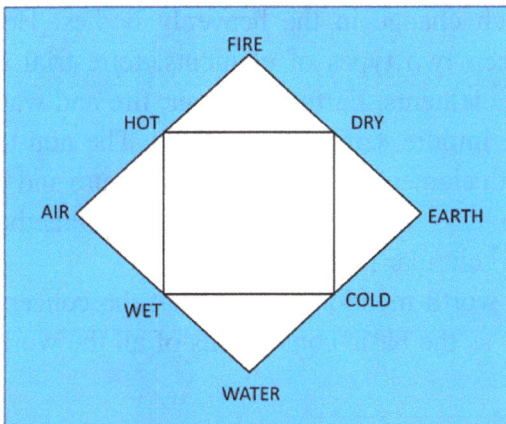

Figure 2.2. Relation between four elements and four forms as envisaged by Aristotle.

fundamental elements. Later, in this book, I will discuss the fundamental elements in details. However, it may be pointed out that the ancient Greeks may not be completely wrong and there are some elements of truth in their wisdom. Modern science tells us that matter can exist in four phases, the solid phase, the liquid phase, the gaseous phase and the plasma phase. If we consider the possibility of a different nomenclature in those times, the four elements can be identified with the four phases of matter as we know now:

$$Land\ (Earth) \leftrightarrow Solid,$$
$$Water \qquad \leftrightarrow Liquid,$$
$$Air \qquad\quad \leftrightarrow Gas,$$
$$and,\ Fire \quad\ \leftrightarrow Plasma.$$

Aristotle also thought that each element has a natural motion: earth (land) and water a downward motion to the center of the world, air and fire an upward motion away from the center of the world. For him the center of the world was Earth. On Earth Aristotle could see change, decay and death. A child grows to adulthood, lives his life and then dies; a flower blossoms and wither away; even a stone shows sign of wear and tear. He then surmised that these elements are impure or corruptible, capable of change. He could not detect any such change in the heavenly bodies. He then distinguished between two types of elements, terrestrial elements and non-terrestrial elements. Earth (land), air, fire and water are terrestrial elements, impure, capable of change. The non-terrestrial element is the fifth element — aether, which is pure and incorruptible. As opposed to terrestrial elements, Aristotle said aether could have only perpetual[3] circular motion.

It may be worth mentioning here that the concept of five elements or roots as the basic constituents of all the worldly materials

---

[3]A motion that continues indefinitely without any external source of energy is called perpetual motion. Physically, it is impossible to have such a motion.

was not particular to Greece only. It was an age-old quest of mankind to understand our complex and diverse Universe in simplest possible terms. The quest led to the philosophy of "reductionism" — that a complex system is nothing but a sum of its parts or constituents. A complex system is known if properties of the constituents are known. It may be appropriate to mention here that reductionism does not contradict the possibility of "emerging" phenomena, i.e. the complex system may possess some property/properties of "wholeness," that is beyond the constituent's properties. Ancient mankind, irrespective of its geographical location, developed this philosophy. The philosophy of reductionism gave birth to the concept of five elements at Greece; in India, the concept of "*Pancha bhoota*" or five elements — land, water, air, fire and sky. China, another ancient civilization, also reduced the Universe to five elements: fire, earth, metal (literally gold), water and wood.

Aristotle also had definite ideas about our Cosmos. The literary meaning of "Cosmos" is Universe seen as a well-ordered whole. He gave a model for Cosmos or Universe. A schematic diagram of Aristotle's Universe is shown in Figure 2.3. Aristotle's Universe was simple; it consisted of Earth, Sun, Moon, five planets (Mercury, Venus, Mars, Jupiter and Saturn) and fixed stars. He gave Earth the

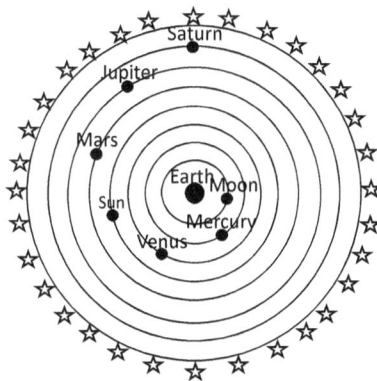

Figure 2.3. Schematic diagram of a geocentric model of Universe as envisaged by Aristotle. The diagram is not to scale.

preferential position, the centre of the Universe. The choice was natural from his point of view; he could see that the Sun and planets revolved around the Earth each day. However, he could not see any perceptible motion of the Earth. In Aristotle's model of Universe, stationary Earth is at the center, and heavenly bodies — spherically shaped Sun, Moon, planets and fixed stars — are embedded in concentric crystalline spheres made of pure, incorruptible aether, executing perfect circular motion around the Earth. The spheres had to be of crystal to account for the visibility of other planets and the stars through it. But what caused the crystalline spheres to move? Aristotle believed that the final cause of all movements was a "Prime Mover." "Movement" for Aristotle was not just the motion of a body from here to there, but much more than that. Movement also included change, growth, melting, cooling, heating etc. Prime Mover, itself unmoved, was not an efficient cause, but the final cause of a movement. In other words, it was the teleological end of a movement. In his model of Universe, Aristotle then conceived an outermost sphere that was the domain of the Prime Mover. The Prime Mover caused the outermost sphere to rotate at a constant angular velocity, which was imparted from sphere to sphere, thus causing the whole thing to rotate. What was this Prime Mover? Aristotle defined it as eternal, incapable of change, most pure as well as most beautiful. In his book *Metaphysics* Aristotle linked the Prime Mover with God and concluded that God is "*a living being, eternal, most good, so that life and duration continuous and eternal belong to God; for this is God.*" The ancient Greeks believed in polytheism; they worshipped multiple Gods. Aristotle was also a polytheist. Yet he conceived the Prime Mover, which many historians believe was the precursor to the Christian God.

2.2 How big is our Earth?

If the spherical shape of the Earth is difficult for a man to comprehend, it is equally difficult for him to comprehend how big it is. Yet, the ancient Greeks had a very good idea about the sizes of the heavenly bodies like Earth, Sun and Moon and also their distances from each other. They measured them as early as 300 BCE! Simple and clear understanding of the physics of shadow formation, eclipse, phases of the Moon and simple geometry enabled them to devise ingenious methods to measure the sizes and distances of the astronomical bodies like Moon, Earth and Sun. The accomplishments may seem trivial now, but considering the great sizes of the astronomical bodies (for example, on the average, circumferences of Sun, Earth and Moon are 4,375,243 km, 40,075 km, and 10,920 km respectively), the great distances involved (approximately 384,440 km lunar distance, and approximately 150 million km solar distance) and the state of science and technology at that time, the accomplishments speak voluminously of their intellect. Indeed, even today, a large number of educated men will be at a loss if asked to devise a simple method to measure the size, say of the Earth.

In *On the Heavens*, Aristotle commented that *"mathematicians who try to calculate the size of the Earth's circumference arrive at the figure 400,000 stades."*[4] He also commented that *"compared with the stars it is not of great size."* He arrived at this remarkable conclusion by observing that,

> "...there are some stars seen in Egypt and in the neighbourhood of Cyprus which are not seen in the northerly regions; and stars, which in the north are never beyond the range of observation, in those regions rise and set. All of which goes to show not only that the Earth is circular in shape, but also that it is a sphere of no great size: for otherwise the effect of so slight a change of place would not be quickly apparent."

Aristotle did not elaborate how the mathematicians arrived at the number 400,000 stades as Earth's circumference. We now know that the Earth's circumference measures 40,075 km and the quoted value 400,000 stades (approximately 74,000 km) overestimated it nearly by a factor of two. Rather accurate measurement of Earth's

---

[4] Greeks measured length in unit of "stade." One stade is approximately 185 meters.

circumference had to wait for about a hundred years from Aristotle's time. In 240 BCE, Eratosthenes of Cyrene, the Greek astronomer, geographer and philosopher, performed a simple but novel experiment to determine Earth's circumference which was accurate within 15%. Eratosthenes (ca. 276–194 BCE) had the best education then available at Greece. After studying in Cyrene and Athens, he moved to the great city of Alexandria in Egypt, and later became librarian of its great library, The Royal Library of Alexandria.[5] Eratosthenes can be considered as the father of geography. Geography is the science of the study of the lands, the features, the inhabitants, and the phenomena of Earth. He wrote a comprehensive treatise about the world, called *Geography*. He was the first person to use the term "geography," which was coined from the two Greek words "*Ge*" meaning the earth and "*Graphia*" meaning writing or a field of study. His experiment on the determination of Earth's circumference is a scintillating example of what a man of intellect can achieve with simple implements and knowledge.

Eratosthenes knew that at the summer solstice (the longest day, about 21 June in the northern hemisphere and 22 December in the southern hemisphere) at noon, the Sun shone directly into a well at Syene (now Aswan), some 781 km south of Alexandria. He then vertically struck a stick of known length (BC) at Alexandria (see Figure 2.4) and at noon, measured the length of its shadow (AC). Eratosthenes was well versed in geometry. A little bit of geometry tells us that the stick makes a right angle[6] with the ground

---

[5]The Royal Library of Alexandria was one of the largest and most significant libraries of the ancient world. Ptolemy I, a Macedonian general under Alexander the Great, and later ruler of Egypt, founded the library in 3rd century BEC. The library was later burnt by Caeser's soldiers.

[6]An angle is the inclination of one straight line to another. If one straight line stands upright on another, then it is a right angle. An angle is measured in units of degree. One complete rotation makes 360 degrees. A right angle corresponds to ¼ of rotation and measures 90 degrees. In mathematics, an angle is represented by the symbol ∠.

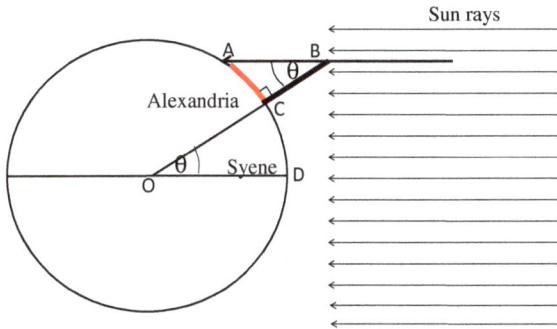

Figure 2.4. Schematic diagram for Eratosthenes' Earth's radius measurement (not to scale).

and the triangle ABC is a right angled triangle.[7] The inclination angle θ of Sun's rays can be determined easily using the relation;

$$\tan \theta = \frac{\text{Length of the shadow}}{\text{Length of the stick}} = \frac{AC}{BC}$$

Eratosthenes measured the angle of inclination $\theta = 7.2°$. Since Sun's rays are parallel,[8] he correctly deduced that the stick at Alexandria, if extended inside the earth, it will make the same angle with the line joining centre of the Earth and Syene. In other words, the part of the Earth between Syene to Alexandria subtends an angle $7.2°$ at the centre of the Earth. He then argued that $360°$ corresponds to Earth's circumference (2π times Earth's radius). From the known distance between Alexandria and Syene, he calculated the Earth's

---

[7]A Triangle is a figure formed by three straight lines joined end to end. It has three angles, hence the name. If one of the angles is a right angle, it is called right angled triangle.

[8]Two lines are parallel when produced to infinity they never meet one another. When two parallel lines are crossed by another, it produced same interior angles e.g. $\angle ABO$ and $\angle BOD$ in Figure 2.4.

circumference to be 787 x 360/7.2 = 39,350 km, closely agreeing with present day measurements.

The Greeks also used simple physics and mathematics to measure the size of the Moon. Aristarchus of Samos (ca. 310–250 BCE) was the first person to propose a heliocentric or Sun-centered Universe. We will talk about it later. In his book, *On the Sizes and Distances of the Sun and Moon*, he detailed the method of measurement. He understood that lunar eclipse occur due to the Moon entering the Earth's shadow. If Sun's rays are parallel then Earth's shadow is a measure of Earth's diameter (see Figure 2.5). He estimated the time Moon took to completely enter into the shadow and the time it was inside the shadow. He found both times to be the same. Rightly, he then concluded that Moon's diameter must be half of the Earth's diameter. Aristarchus was off by a factor of two, Earth is actually 4 times bigger than Moon. The reason is now understood. While it is reasonable to assume Sun's rays are parallel on a small portion of Earth (as Eratosthenes did) it is certainly wrong when the entire Earth is considered. When Earth is considered as a whole, Sun's rays are not parallel and Earth's shadow will have a conical shape (see Figure 2.5) and is not an exact measure of Earth's diameter.

Aristarchus of Samos could also measure the ratio of solar (Sun–Earth) distance and lunar (Moon–Earth) distance. He correctly deduced that at quarter moon, the three celestial bodies,

Figure 2.5.   Moon's size measurement during lunar eclipse (not to scale). Left figure shows Aristarchus' concept of lunar eclipse. Right figure shows the present understanding.

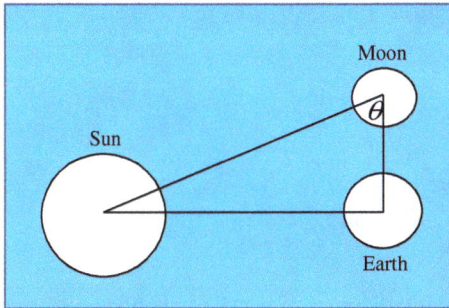

Figure 2.6. Schematic diagram for Sun, Earth and Moon positions at quarter moon (not to scale).

Earth, Sun and Moon make a right angled triangle (see Figure 2.6). If the angular difference between the Sun and Moon is measured, then the simple relation,

$$\tan\theta = \frac{\text{Earth–Sun distance}}{\text{Earth–Moon distance}},$$

determines the ratio of the distances. Aristarchus could not measure the angle very accurately. While his method was correct he arrived at a wrong answer that from Earth, Sun was at 18–20 times larger distance than the Moon (actual factor being around 400).

The first measurement of lunar distance was made by the Greek astronomer and mathematician, Hipparchus of Nicaea. Very little is known about Hipparchus apart from that he has been a working astronomer at least from 162 to 127 BCE. He used the parallax method to make the measurement. Parallax is the apparent shifting of a target position due to change in the observer's position. The effect is easy to understand. Many of us have experienced the effect. Here is a simple experiment. Stretch one of your hands and lift the thumb. Close one eye and align the thumb with a distant object. Now, keeping your arm fixed, switch the eye (i.e. see the object with the other eye). The object is shifted with respect to your thumb. This is the parallax effect.

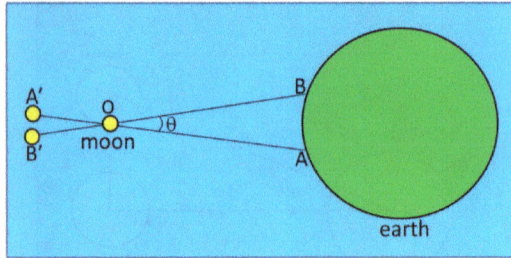

Figure 2.7.   Schematic diagram of parallax method employed by Hipparchus to estimate Moon–Earth distance (not to scale).

The application of parallax method to measure the lunar distance is explained in Figure 2.7. Say, the Moon is at the position O. From the position A on the Earth, the Moon will appear to be located at A′ but from the position B at B′. The apparent shifting of Moon's position by the angle θ can be measured. The triangle OAB can be assumed to be an isosceles triangle (i.e. two sides are equal). If the distance between A and B is 2S, the Moon's distance D from the surface of the Earth can be obtained directly,

$$\tan\frac{\theta}{2} = \frac{S}{D}.$$

In reality, Hipparchus estimated the distance from observing full and partial solar eclipses. He attributed the difference entirely to the Moon's observable parallax effect. From his calculations, Hipparchus estimated the Moon's mean distance from the Earth is approximately 63 times the Earth's radius. (The true value is about 60 times.)

Once the size of Moon and the ratio of solar to lunar distances were known, it was not difficult for the Greek astronomers to infer about the size of the Sun. They knew that during the solar eclipse, Sun was completely blocked out by the Moon. Sun and Moon must have the same angular size. The geometrical arrangement during a

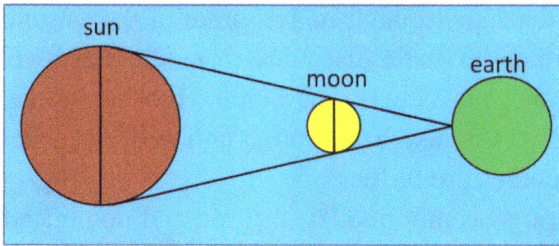

Figure 2.8. Schematic diagram for positions of Sun, Moon and Earth during solar eclipse (not to scale).

solar eclipse (see Figure 2.8) then allows one to use properties of similar[9] triangle,

$$\frac{\text{Solar (Earth--Sun) distance}}{\text{Lunar (Earth--Moon) distance}} = \frac{\text{Diameter of Sun}}{\text{Diameter of Moon}},$$

and extract the diameter of the Sun. There are several more ingenious measurements by the ancient Greek astronomers I have not discussed here. The measurements were not always very precise; however, they demonstrate their deep understanding of geometry, trigonometry and processes like eclipse, phases of the moon etc.

2.3 Earth is center of the Universe

In his 340 BCE book *On the Heavens* Aristotle argued for a geocentric or Earth-centered model for the Universe. Aristotle's Universe was simple; it consisted of fixed Earth at the center, surrounded by the Moon, Mercury, Venus, Sun, Mars, Jupiter, Saturn and fixed stars, embedded in crystalline spheres, rotating around the Earth. The spheres had to be of crystal to account for the visibility of other

---

[9] If two angles of a triangle have measures equal to the measures of two angles of another triangle, then the triangles are similar.

planets and the stars through it. The order of the astronomical bodies is according to their distances from the Earth. Even before Aristotle's time, the Greeks knew that Moon is the nearest astronomical body. They also had a rough notion of distances; the slowest moving bodies are further off.

Aristotle's geocentric model survived and dominated until 1600 CE when Copernicus proposed his heliocentric or Sun-centered model. However, very early in Greek history there were proposals for a non-Geocentric cosmic model. Philolaus of Croton (470–385 BCE) was a Greek philosopher, a contemporary of Socrates. He wrote a book *On Nature*, few fragments of which are still preserved. He argued that the Cosmos and everything in it were made up of two basic types of things, limiters and unlimiteds. Furthermore, limiters and unlimiteds were not combined in a haphazard way but were subject to a "fitting together" or "harmony." He wrote,

> "Nature in the world-order was fitted together both out of things with are unlimited and out of things with are limiting, both the world-order as a whole and everything in it."

Philolaus was remarkably reticent about what he meant by limiters and unlimiteds. The modern interpretation is that unlimiteds are continua undefined by any structure or quantity: material elements such as earth, air, fire and water but also space and time. Limiters set limits in such unlimiteds and include shapes and other structural principles.

In Philolaus' model of Universe in ten concentric circles, heavenly bodies, five planets, Sun, Moon, Earth, fixed stars and a mysterious "counter-Earth", moved around a central fire. Both the central fire and the counter-Earth were not visible from the Earth. Earth in Philolaus' model was not spherical, rather a flat Earth, inhabited part always facing the Sun. The arrangement of Earth, counter-Earth, Sun and central fire is schematically shown in Figure 2.9. The central fire in the model was the driving force moving the

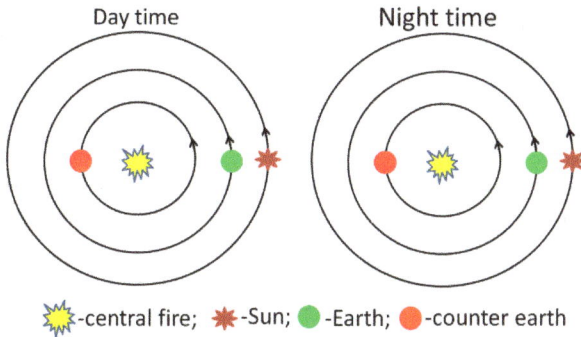

Figure 2.9. Arrangement of the central fire, Sun, Earth and counter-Earth in Philolaus' non-geocentric model of Universe.

heavenly bodies. The tenth body, the counter-Earth was introduced for a numerological reason, to make the total number of heavenly bodies as 10. Philolaus considered 10 to be a holy number.

Philolaus of Croton believed in Pythagorean doctrine. Pythagoras of Samos (ca. 570–495 BC) was a great mathematician. We are well acquainted with the famous Pythagoras theorem.[10] Born in a wealthy merchant family, he had the best education and was taught by the great Greek philosophers — Thales[11] and Anaximander.[12] He travelled extensively and learned from Egyptian priests, Indian sages. Later, he established a school at Croton. His

---

[10] Pythagoras theorem relates three sides of a right angled triangle — the square of the hypotenuse (the side opposite to the right angle) is equal to the sum of squares of the other two sides.

[11] Thales of Miletus (ca. 624–546 BCE) was a pre-Socratic Greek philosopher and mathematician. He took the holistic view that origin of everything in nature is the single material substance — water. Everything was made of water and ultimately perished into water.

[12] Anaximander (ca. 610–546 BCE) was a pre-Socratic Ionian philosopher. A student of Thales and teacher of Pythagoras, Anaximander believed that everything in nature followed a rule of law and made the first comprehensive attempt to explain the origins both of man and the Cosmos.

mathematical-philosophical teachings gave birth to the religious-mathematical cult called Pythagoreans. The cult survived long after his death and over the years Pythagoras became a subject of elaborate legend, he was equated with divinity (legend was that he possessed golden thigh) with supernatural power. Pythagorean sects were governed by a set of rules, e.g. no worldly possession, plain attire, strict vegetarianism (beans were forbidden), communal living, secret rites, etc. The overriding dictum of Pythagorean school was harmony and number. Apart from being a great mathematician, Pythagoras was also a fine musician and used to play lyre.[13] He noticed that the vibrating strings produced harmonious tones when the ratios of the lengths of the strings were whole numbers and that these ratios could be extended to other instruments. Pythagoreans believed that numbers were not mere physical abstraction, but something real which existed in and composed all things. Pythagorean attitude to numbers was exemplified by Philolaus,

> "And indeed all things that are known have number. For it is not possible that anything whatsoever be understood or known without this."

For Pythagoreans, each number has a character and meaning. For example, the number "one" was the generator of all numbers; "two" represented opinion; "three" harmony; "four" justice; etc. The holiest number is 10, as it contains the first four integers ($10 = 1 + 2 + 3 + 4$). The mystical figure tetractys (see Figure 2.10), which Pythagoreans used in their religious rites, also contains 10 points. In a sense, the special place for the number 10 from an abstract mathematical argument rather than from something as mundane as counting the fingers on two hands speaks of the intellectual achievement of Pythagoras.

---

[13] A lyre is a string instrument used in ancient Greek. It is similar in appearance to a small harp but with distinct differences.

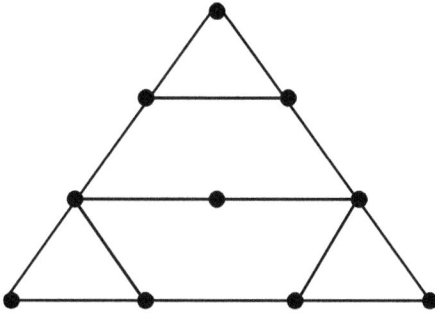

Figure 2.10. The mystical figure of tetractys, used in secret rites by Pythagoreans. The figure was believed to represent the organization of space. The first row represents zero dimensions (a point), the second row one dimension (a line of two points), the third row two dimensions (a plane defined by a triangle of three points) and the fourth row three dimensions (a tetrahedron defined by four points).

In his non-geocentric cosmic model, Philolaus put a central fire around which the heavenly bodies rotated. Aristarchus of Samos (ca. 310–250 BCE), the first person to measure sizes of Sun and Moon and their distances, also proposed a non-geocentric model of cosmos. He built on Philolaus' model by replacing the "central fire" by the Sun. He reasoned that since the Sun and the Moon had the same angular size (Sun was completely obscured by Moon during solar eclipse), but the Sun was 20 times further (or so he thought), then the Sun must be 20 times bigger than the Moon. The comparatively large size of Sun led him to suggest the first Sun-centered (or heliocentric) model for the Universe.

Even though the Greeks for two centuries built on the heliocentric model, Aristarchus' view was not accepted. First of all, it was contrary to the visible world, we do not experience motion of the Earth. Moreover, Aristotelians argued that in a heliocentric model, one would have observed parallax effect of the stars; if the Earth rotated around the Sun then measured at six months interval, star's position would change. For example, in Figure 2.11, from Earth's

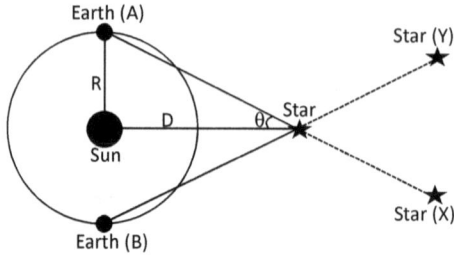

Figure 2.11. Schematic diagram explaining the parallax method to determine the distance of a nearby star.

position at A, the star will appear to be at X. Six month later, Earth will be at the position B. From position B, the same star will appear to be at Y. The Greeks could not see any such shift in a star's position. They did not realize that the stars were very far away and the shift in position was too small to be detected by the naked eye. Nowadays, the parallax method is used to measure the distance of not-too-distant stars. Simple geometry will make you understand that the star's distance can be obtained as,

$$\tan \theta = \frac{\text{Sun–Earth distance}}{\text{Sun–Star distance}} = \frac{R}{D}.$$

2.4 Circle within a circle

The ancient Greeks rejected Aristarchus' model and built on Aristotle's proposal for the geocentric (Earth-centered) model for the Universe. The essentials of a geocentric model are, (i) Earth is at the center of the Universe, (ii) heavenly bodies (Sun, planets and Moon) are attached to different celestial spheres and move in uniform, circular motion, and (iii) fixed stars are at the outermost celestial sphere. It is also the abode of Gods. A schematic diagram for a geocentric model is shown in Figure 2.3.

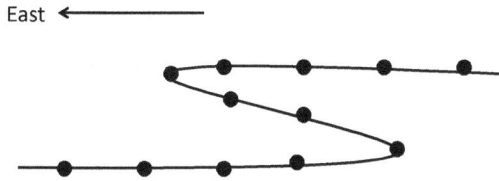

Figure 2.12. Pictorial depiction of retrograde motion of planets. The black dots are planet positions.

However, Aristotle's geocentric model faced a difficulty: it could not explain one of the observed features of the planetary motion — the retrograde motion. When observed from the Earth some of the planets appear to undergo a curious motion. The planet moves towards the east, stops and suddenly reverses its direction of motion, continues its motion in the reverse direction for a certain time and then the direction of motion reverses again. A schematic of the planetary retrograde motion is shown in Figure 2.12. In a geocentric model, with the Earth at the center and planets revolving around the Earth in circular orbits, one cannot explain the retrograde motion. The planets can have only unidirectional motion.

Claudius Ptolemy (90–168 CE), the Egyptian astronomer, mathematician, and geographer, improved the geocentric model to explain the retrograde motion. Very little is known about his life apart from what can be inferred from his writings. As the name suggests, he was a Roman subject of Greek descent. He lived in Alexandria. His geocentric model, which he expounded in the great astronomical treatise, *Almagest*, dominated the world's astronomy till 16th century. Originally, the treatise was titled *Syntaxis Mathematica*. Muslim scholars translated it into Arabic under the title *al-majisti* literally meaning the greatest or the best. The English name *Almagest* was derived from the Arabic name. In *Almagest*, Ptolemy presented his astronomical model in convenient

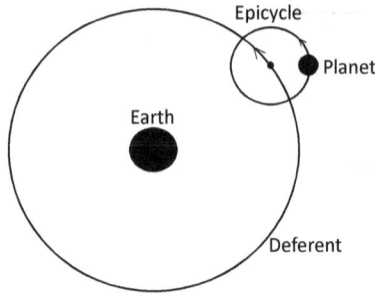

Figure 2.13. Ptolemy's ingenious method for explanation of the retrograde motion.

tables, which could be used to compute the future or past positions of the planets. It was a must for all the astronomers.

To explain the retrograde motion, Ptolemy devised an ingenious method. Schematically, it is shown in Figure 2.13. Each planet moves on a small circle, called an epicycle. The center of the epicycle moves on a larger circle, called deferent. With the introduction of epicycles, the model can explain the retrograde motion. If Earth is at the center of the deferent then while the center of epicycle moves eastward with respect to the earth, the planet can now have both eastward as well as westward motion. However, detailed calculations of planetary positions required that the Earth could not be the center of the deferent. Ptolemy introduced an imaginary point "equant." The center of the deferent is the midpoint between the Earth and the imaginary point "equant." With respect to the equant, the center of the epicycle always moves at the same speed.

Ptolemy was unhappy about introducing the imaginary point equant. It sacrificed one of the long cherished Greek principles, namely the uniform circular motion of planets around the Earth. However, he was forced to keep it to conform to the observed planetary motions. Ptolemy was a scientist and understood the supremacy of observational truth. He understood that primary objective of any scientific modeling is to explain the observational truths and if required, to predict for future observations. Ptolemy's model was rather complex, each planet requiring a different equant, sometimes with multiple epicycles. Reportedly, King Alfonso X[14] of Castile, learning about Ptolemy's model commented,

> "If the Lord Almighty had consulted me before embarking on creation thus, I should have recommended something simpler."

Ptolemy's geocentric model was unchallenged until the 16th century. The model was acceptable to the then all powerful Church. The model has ample scope for divine intervention. The outer celestial sphere where Ptolemy kept his fixed stars can be the abode of the divinity. The model was also practical; it did correctly predict movements of the planets (occasionally requiring an adjustment in planetary positions). How good the model was can be understood if we consider that modern planetariums, built using gears and motors, essentially reproduce the Ptolemaic model for the appearance of the sky as viewed from a stationary Earth. It is no wonder that Ptolemy's model survived so long.

---

[14] King Alfonso X (1221–1284), also known as Alfonso the wise, was King of Castile, a powerful state in Iberian peninsula. Himself a scholar, he had many scholars in his court, and he actively participated in their writing and editing.

Chapter 3

Early Modern Cosmology

> The diversity of the phenomena of nature is so great, and the treasures hidden in the heavens so rich, precisely in order that the human mind shall never be lacking in fresh nourishment.
>
> Johannes Kepler

3.1 European dark age and Hindu and Muslim astronomy

The Middle Ages of the world's history, the period 500–1500 CE, is generally called the Dark Age of Europe. In that period development of new scientific theories in Europe came almost to a halt. The primary reason for the decline in new scientific ideas was the disintegration of the Roman Empire, or more specifically the Western Roman Empire. Roman Empire was at its peak around 100 CE, comprising a vast contiguous territory throughout Europe, North Africa, and the Middle East. In 285 CE, for administrative reasons, the vast empire was divided into a western and eastern part. Around 500 CE, the western part of the Roman Empire started to disintegrate due to internal conflict, and invasion from Germanic barbarians. In the ensuing period of social, political and economic unrest science could not flourish, literacy dropped, centers of higher learning shifted to monastic and Catholic schools with *Bible* being the major study material. Overall, European mental efforts were directed towards non-scientific pursuits. Irrational theories gradually engulfed the whole of science. Astrology challenged astronomy, alchemy infiltrated natural science and magic insinuated itself into medicine. The eastern part of the Roman Empire, also called Byzantine Empire, however, survived till 1500 CE, until it fell to Turkish invasion. However, it will be wrong to assume that scientific enquiries came to a stop in the Middle Ages. Though scholastic endeavors of monastic Catholic schools mostly concerned *Bible*, the monks, to care for the sick also studied medicine. They also studied astronomy. To fix the dates for religious ceremonies, they had to observe the stars. Their astronomy kept alive mathematics and geometry.

Towards the end of 12th century, Western Europe slowly crawled out of the Dark Age. Europe went through a major social, political and economic transformation, revitalizing itself intellectually. The Muslim scholars had translated many of the ancient Greek texts into Arabic. They were translated back into Latin. Indeed, in 11–12th century, it was common for European scholars to flock Spain to learn Arabic and translate books from Arabic into Latin. Slowly, learning centers shifted back to universities from monastic Catholic schools. The process culminated in 15th century Italian Renaissance and finally to the scientific development of the 17th century.

While Europe was suffering through the Dark Age, Indian or Hindu astronomy was making great progress. Hindu astronomy has a long history. One of the earliest written documents on astronomy is *Vedanga Jyotisha* of sage Lagadha. As the name suggest, *Vedanga* literarily means "limbs of the Veda". There are six Vedangas: (i) *Shiksha* or phonetics or pronunciation, (ii) *Kalpa* or ritual, (iii) *Vyakarana* or grammar, (iv) *Nirukta* or etymology, (v) *Chandas* or meter and (vi) *Jyotisha* or astronomy. The basic purpose of Vedangas was to facilitate the study and understanding of the four sacred books of Hindus, *Rig Veda*, *Sama Veda*, *Yajur Veda* and *Atharva Veda*. Though, in modern times, *jyotisha* is generally translated as "astrology," in the Vedic period, it meant astronomy. Jyotisha or astronomy enjoyed high esteem in ancient India. Lagadha wrote,

> "Like the comb on the head of a peacock or a jewel on the head of a snake, this scientific composition of Jyotisha (Vedic astronomy) occupies the position at the top of all (six) annexures of Vedas."

The exact date of *Vedanga Jyotisha* is disputed. Scholars fix the time between 850–1370 BCE though the extant text is dated around the final centuries BCE. *Vedanga Jyotisha* came in two recensions, one belonging to the *Rig Veda* and the other belonging to *Yajur Veda*, the basic content of both being almost the same. Lagadha stated explicitly the very purpose of *Vedanga Jyotisha*:

> "The Vedas have indeed been revealed for the sake of the perfor-mance of the sacrifices. But these sacrifices are dependent on the (various segments of) time. Therefore, only he who know the lore of time, viz. Jyotisa, understands the performance of the sacrifices (fully)."

Vedanga Jyotisha is essentially a handbook for determination of time for sacrifices (*yajnas*), rituals and allied processes. A distin-guishing feature of Vedic astronomy is the "Asterism" or the "Nakshatra system." Vedic astronomers used sidereal time. The word sidereal originated from the Latin "*sidus*" meaning star or group of stars. Sidereal year is the time taken for the Sun to return to the same position with respect to the fixed stars, as viewed from Earth. The Greeks, on the other hand, used the tropical year. The apparent path of the Sun, in the background of the celestial sphere,[1] is called the ecliptic. Earth's equatorial plane projected on the

---

[1] Celestial sphere is an imaginary sphere, concentric with the Earth. The heavenly bodies are so remote that they can be thought to be moving on the inside surface of the celestial sphere. The sky as we see it is the upper dome of the celestial sphere.

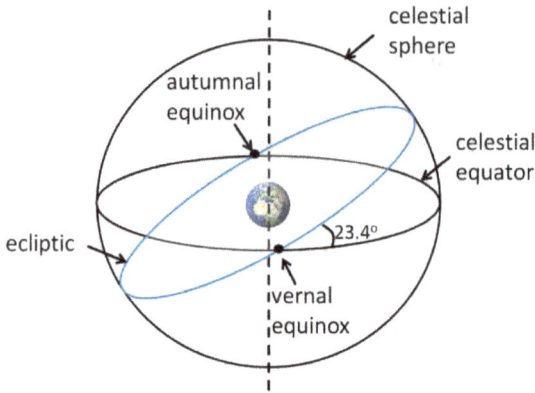

Figure 3.1. Schematic representation of the celestial sphere, ecliptic (apparent path of the Sun) and celestial equator.

celestial sphere is called the celestial equator and since Earth's orbital plane is tilted with respect to the equatorial plane, it bisects the ecliptic at two points, known as vernal equinox and autumnal equinox (see Figure 3.1). The time taken by the Sun to move from vernal equinox to vernal equinox is called a "tropical" year. Due to precision of equinox,[2] a sidereal year differs from the tropical year. It is longer approximately by 20 minutes.

In Greek astronomy the ecliptic is called zodiac, literarily meaning "circle of little animals." Along the ecliptic, 30 degrees apart, Greeks imagined 12 pictures of animals, which became some of the first constellations.[3] They are Aries, Taurus, Gemini, Cancer, Leo, Virgo, Libra, Scorpio, Sagittarius, Capricorn, Aquarius, and Pisces. Interestingly, the only inanimate sign among the twelve is Libra,

---

[2] In astronomy, axial precession is a gravity-induced, slow, and continuous change in the orientation of an astronomical body's rotational axis. Traditionally, Earth's axial precession is called precession of the equinoxes. Equinoxes are the dates or times when day and night are of equal length. It occurs twice a year, around 21 March and 23 September.

[3] A constellation is a group of stars that, when seen from Earth, form a pattern.

which means balance. In Greek mythology, Astraea was the Goddess of Justice. She was the last of the Gods to flee the Earth as the mankind became wicked. Astraea ascended to the heaven and became the constellation Virgo. The scales of justice she carried became the nearby constellation Libra. Originally, the Greeks saw the scales of balance as being the claws of the scorpion, hence the name.

Truly speaking, zodiac signs are nothing more than a coordinate system, more specifically the "ecliptic coordinate system." The ecliptic coordinate system is similar to the latitude–longitude coordinate system (see Figure 3.2) used in locating a position on the Earth's surface. Latitudes are imaginary lines on the Earth's surface, parallel to the equator and are a measure of the location north or south of the equator. Equator (midway between North and South Poles) is assigned 0 degree latitude. The latitude of the North and South Poles are +90 degrees and –90 degrees respectively. Longitudes are imaginary lines from the North Pole to the South Pole and measure the location east or west of the prime meridian, the longitude which passes through Greenwich, England. The combination of

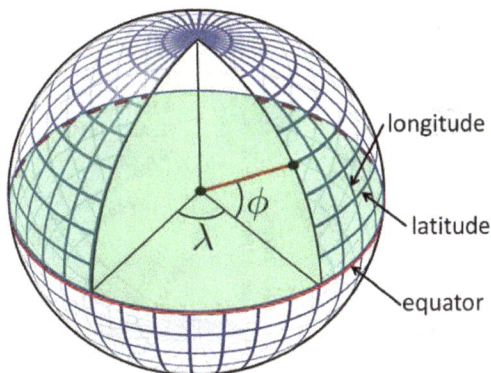

Figure 3.2. Latitude and longitude coordinate system to pinpoint a location of Earth's surface.

latitudes and longitudes establishes a grid by which location of a place can be pinpointed. For example, the location 22.57° North, 88.37° East, identifies a place at 22.57° north of the equator and 88.37° east of Greenwich. It is the city of Joy, Kolkata.

The celestial sphere can also be imagined to be crisscrossed by the lines of latitudes and longitudes. In ecliptic coordinate system, the ecliptic is assigned zero degrees latitude and the position of the Sun at the vernal equinox is zero degrees longitude. Traditionally the zodiac sign Aries is associated with the vernal equinox. The Greek division of year by zodiac signs is shown in Figure 3.3.

While in Greek astronomy the ecliptic is divided into 12 divisions, in Hindu astronomy, the ecliptic is divided into 27 or 28 divisions called *nakshatras*. The number of nakshatras reflects the number of days in a sidereal month. The sidereal month is the time taken by the Moon to return to the same position with respect to the fixed stars (modern value: 27.32 days). Each day, the Moon traverses the width of

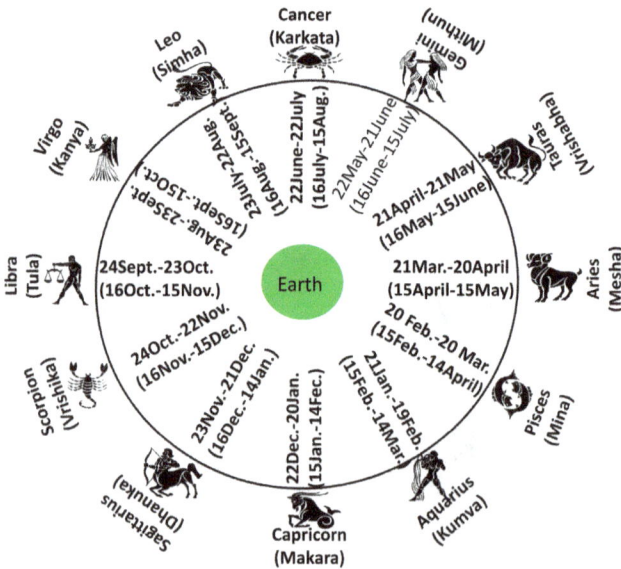

Figure 3.3. Zodiac signs (and Hindu *rashis*) along the ecliptic are shown.

a nakshatra. For ease of calculation, in *Vedanga Jyotisha*, Lagadha used 27 nakshatras, each covering 13°20' of the ecliptic. The 27 nakhastras are: *1. Ashvinee, 2. Bharanee, 3. Krittika, 4. Rohinee, 5. Mrigasheera, 6. Aadraa, 7. Punarvasu, 8. Pushya, 9. Aashleshaa, 10. Maghaa, 11. Poorvaa Phalgunee, 12. Uttaraa Phalgunee, 13. Hasta, 14. Chitraa, 15. Swaatee, 16. Vishakhaa, 17. Anuraadhaa, 18. Jyeshthaa, 19. Moola, 20. Poorvaashaadha, 21. Uttaraashaadhaa, 22. Shraavana, 23. Shravisthaa, 24. Shatabhisha, 25. Poorva Bhaadrapadaa, 26. Uttaraa Bhaadrapadaa, 27. Revatee.* The 28th nakshtra, used prior to Lagadha, is called *Abhijit*.

It may be mentioned here that much later, under the influence of Hellenistic Greeks, Hindus introduced the zodiac system. It was called *"Rashi."* Naturally, the Hindu and Greek names differ; the symbols however, remained the same. The Hindu zodiacs or rashis are also shown in Figure 3.3. The Hindus use sidereal year, and as shown in Figure 3.3, the zodiac signs fall in different times of the year.

A disconcerting feature of *Vedanga Jyotisha* is the complete absence of planets. In Sanskrit, a planet is called *graha*. Throughout *Vedanga Jyotisha*, no mention of any *graha* can be found. Yet, in *Rig Veda* there are several verses dedicated to *Vrihaspati*, the Hindu name for the planet Jupiter. However, in Hindu sacred literature, Vrihaspati is also called *Guru*, the teacher. It is then uncertain whether Vrihaspati in *Rig Veda* refers to the planet Jupiter or some deity. It is unlikely that ancient Hindus could identify distant, feeble stars, but could not see the much nearer and much brighter planets. One possible reason for the absence of planets in Vedic astronomy is that they play no role in the determination of time. As mentioned earlier, *Vedanga Jyotisha* was essentially a handbook for calculating times for rituals, sacrifices etc. Possibly, Lagadha did not want to encumber his treatise with redundant information.

Another distinguishing feature of *Vedanga Jyotisha* is rather a small duration of *yuga*. Yuga, as a measure of time, is a Hindu concept. In *Vedanga Jyotisha*, Lagadha used 5 solar years as the measure of yuga. The following table gives the division of time in *Vedanga Jyotisha*:

10 matras   = 1 kastha
125 kasthas   = 1 kala
10 1/20 kalas = 1 nadika
2 nadikas   = 1 muhurta
30 muhurtas = 1 day (i.e. the civil day)
366 days   = 12 solar months or 1 solar year
5 solar years = 1 yuga

A 5-year yuga is much shorter than the yuga used by later astronomers like Aryabhata. Aryabhata was the most important post-Vedic Hindu mathematician and astronomer. He was born in 476 CE in Asmaka (now in Maharashtra), one of the 16 Mahajanapada (great country) mentioned in Buddhist texts. He lived in Kusumpura, i.e. Pataliputra or modern Patna, during the reign of Buddhagupta, the last great King of the Gupta dynasty.[4] In 5th century, Gupta King, Kumaragupta founded "*Nalanda,*" the great Indian learning center from 5th century to 12th century. It is reasonable to believe that Aryabhata was the Kulapati (Head of University) of Nalanda, which was flourishing in 5th and 6th century, when Aryabhata lived.

Aryabhata wrote several mathematical and astronomical treatises, most of which are now lost. His most important work *Aryabhatiya* had survived up to the modern times and was written when he was only 23 years old. It has four chapters: Dasagitika, Ganita, Kalakriya and Gola. Each chapter has several verses or

---

[4] Gupta dynasty was founded by King Sri Gupta around 200 CE. For more than 300 years, the dynasty ruled a large part of the Indian subcontinent. The Gupta age is supposed to be the golden age in Indian history. The peace and prosperity created under the leadership of the Guptas enabled the pursuit of scientific and artistic endeavors.

"sutras." In Dasagitika, Aryabhata defined large units of time, and an astronomy at variant with *Vedanga Jyotisha*. While Lagadha used a 5-year yuga in *Vedanga Jyotisha*, Aryabhata defined yuga as 4,320,000 years. In the second and third verse of Dasagitika, he defined yuga, kalpa (a day of Brahma) and manu as measures of times,

> "In a yuga, the eastward revolutions of the Sun are 4,320,000; of the Moon, 57,753,336; of the Earth, 1,582,237,500; of Saturn, 146,564; of Jupiter, 364,224; of Mars, 2,296,824; of Mercury and Venus, the same as those of the Sun."
>
> A day of Brahma (or a kalpa) is equal to (a period of) 14 manus, and (the period of one) manu is equal to 72 yugas.

We find that by the time of Aryabhata, Indian astronomers were referring to the planets explicitly. In the remaining verses of Dasagitiaka, Aryabhata gave a comprehensive maxim necessary for calculations in astronomy. He also gave a sine table, the first ever constructed in the history of mathematics. The last verse of the chapter say,

> "Knowing this Dasagitika-sutra (giving) the motion of the Earth and planets, on the Celestial Sphere (Sphere of asterism or Bhagola), one attains the Supreme Brahman after piercing through the orbits of the planets and stars."

In Ganita, Aryabhata gave various mathematical formulas, formulas for square root, cubic root, area of a circle, volume of a sphere etc. Importantly, he gave a rather accurate formula for the circumference–diameter ratio of a circle — the value of π.

> "100 plus 4, multiplied by 8, and added to 62,000: this is the nearly approximate measure of the circumference of a circle whose diameter is 20,000."

This gives,

$$\pi = \frac{\text{circumference of a circle}}{\text{diameter of the circle}} = \frac{62{,}832}{20{,}000} = 3.1416.$$

This value did not appear in any earlier work of mathematics and it is an important contribution of Aryabhata, who also knew that the value is only an approximation!

In Kalakriya or the reckoning of time, in various verses, Aryabhata presented a geometrical model for planetary motion, with epicycles of equal linear velocities. In Gola, Aryabhata discussed spherical astronomy. He defined the celestial sphere (called Bhagola), demonstrated the motion of the heavenly bodies, gave various verses to calculate the motion of the Moon and the planets. One verse in Gola is rather controversial:

"Just as a man in a boat moving forward sees the stationary objects (on either side of the river) as moving backward, just so are the stationary stars seen by people at Lanka (at the equator), as moving exactly towards the west."

The natural interpretation of the verse is that an observer at the equator of the Earth, which rotates towards the east, sees stationary celestial objects as though moving westward. It is the first ever mention of Earth's rotation. At that time, rotation of Earth was inconceivable. The verse was severely criticized by later astronomers like Varahamihira (505–587 CE), Brahamagupta (ca. 598–665 CE) etc. Aryabhata himself also appears to be confused. In the next verse, he contradicted himself,

"(It so appears as if) the entire structure of the asterisms together with the planets were moving exact towards the west of Lanka, being constantly driven by the provector wind, to cause there rising and setting."

Table 3.1. Mean motion of planets as calculated in *Aryabhatiya* compared with Ptolemy's calculations and modern values, in days.

	Aryabhata	Ptolemy	Modern value
Sun	365.25858	365.24666	365.25636
Moon	27.32167	27.32167	27.32166
Mercury	87.96988	87.96935	87.9693
Venus	224.69814	224.69890	224.7008
Mars	686.99974	686.94462	686.9797
Jupiter	4332.27217	4330.96064	4332.75637
Saturn	10766.06465	10749.94640	10759.201

What is "provector wind" is not understood, but similarity with Aristotle's prime mover cannot be unnoticed. Overall, in *Aryabhatiya*, Aryabhata presented a rather accurate geocentric model of Universe. Table 3.1 compares Aryabhata's calculation of the mean motion of Sun, Moon and various planets with Ptolemy's calculation and also with their modern values. One can see that his calculations were better than that of the Greek astronomer Ptolemy and very close to the actual values.

In the years following Aryabhata, various astronomers contributed enormously in developing astronomy. Brahmagupta (598–668 CE), Varahamihira (505–587 CE), Bhaskara (ca. 600–680 CE), Lalla (ca. 720–790 CE) and many others perused astronomy and enriched it. While their works perfected or eased calculations of planetary orbits, all of them were confined within the geocentric or Earth-centered model of Universe.

The European Dark Age also coincides with the rise of Islam. In the golden age of Islam, during 800–1500 CE, Muslim scholars made spectacular progress in mathematics and astronomy. An academy or intellectual center at Bagdad called Bayt al-Hikmah or "House of Wisdom," greatly facilitated this progress. House of Wisdom was established by the Caliph Harun al-Rashid (reigned

786–809 CE) but it flourished during his son al-Ma'mun's regime. Renowned 9th century Arab mathematician, al-Khwarizmi studied in the House of Wisdom. From his famous book, *Kitab al-jabr wal-muqabala*, we got the term *algebra*. Caliph al-Ma'mun himself was adept in various branches of science, medicine, philosophy, astronomy etc., and under his patronage, the pursuit of knowledge became a dominant feature of the Caliphate. Scholars were encouraged to translate medieval works and reputed scholars like Hunayn ibn Ishaq[5] would earn in gold of equivalent weight of the manuscript they translated. The vast Greek and Indian literature, pertaining to philosophy, mathematics, natural science, and medicine were translated into the Arabic language. Study of astronomy was of great importance. Ptolemy's *Syntaxis Mathematica* was translated into Arabic and generated a flurry of research activities. A major impetus for the astronomical studies came from the religion itself. A large number of verses in the Muslim Holy Book, the *Quran*, provide information on astronomy. A few examples from the English translation of *Quran*, by Dr. Muhammad Taqi-ud-Din al-Hilali and Dr. Muhammad Muhsin Khan, are given below:

> Sura[6] 7, Ayat 96: (He is the) Cleaver of the day break. He has appointed the night for resting, and the Sun and the Moon for reckoning. Such is the measuring of the All-Mighty, the All-Knowing.

---

[5] Hunayn ibn Ishaq was an influential Arabic scholar, physician and scientist. Apart from translating numerous classical Greek texts into Arabic and Syriac, he made original contributions to medicine and ophthalmology. His ten-volume treatise on physiology, anatomy and treatment on eye was the first systematic investigation on eye diseases and provided the foundation of ophthalmology worldwide.

[6] Quran is divided into 114 *sura* or chapters. Each chapter is further divided into several *ayat* or verses.

Sura 7, Ayat 96: It is He Who has set the stars for you, so that you may guide your course with their help through the darkness of the land and sea. We have (indeed) explained in detail our ayat (proofs, evidences, verses, lessons, signs, revelations, etc.) for a people who know.

Sura 55, Ayat 5: The Sun and the Moon run on their fixed courses (exactly) calculated with measured out stages for each (for reckoning).

The three verses cited above clearly indicate that Islam encouraged the use of the heavenly bodies for fixing time and direction. Religious observances like *Salah* and *Qibla* also called for the study of heaven. *Salah* is the Arabic word for prayer. Muslims are required to pray five times in a day. Timings for the prayers need to be known. *Qibla* is the Arabic word for directions. For believers in Islam, Kaaba, the 60 feet by 60 feet by 60 feet stone building at the center of the Mosque at Mecca, Saudi Arabia, is the most sacred place. All other mosques are required to be oriented in the direction of Kaaba. Also, Muslims are required to pray facing that direction. Both of these require study of heavenly bodies.

Before the 8th century, Muslim astronomy was mainly observational. Assimilation of Greek and Indian knowledge enabled Muslim scholars to add mathematical rigor to their observations. And they then expanded the existing knowledge. Various astronomical terms e.g. Nadir, Zenith are of Arabic origin and so are names of numerous stars. They also greatly improved observational astronomy. The world's first state-sponsored observatory was built by Caliph al-Ma'mun. Muslim scholars improved upon the previous measurements and added new ones. However, all the astronomical activities were confined within the ambit geocentric model of the Universe. While the scholars criticized Ptolemy's model and tried to improve upon it, they did not dare to displace the Earth from the privileged place at the center of the Universe.

3.2 Displacing the Earth from the center

In the 16th century, Ptolemy's geocentric or Earth-centered model was challenged by Nicolaus Copernicus (February 19, 1473–May 24, 1543), the Polish mathematician and astronomer. He was a polyglot[7] and a polymath.[8] As it was the custom of the day, along with mathematics, astronomy, Copernicus also studied Latin, geography, and philosophy at Cracow University. The mathematical skill he demonstrated, in developing the heliocentric or Sun-centered model of the Universe, was acquired during his university study. Upon graduating from Cracow, Copernicus took a canon's[9] position at Torun, one of the oldest cities in Poland. The position gave him the opportunity to continue his interest in astronomy. Later he took leave from his canon's position to study religious law at the University of Bologna. Copernicus also studied medicine at the University of Padua. At the University of Bologna, Copernicus was taught astronomy by the astronomy professor Domenico Maria Novara who influenced his thinking beyond the then prevailing Ptolemy's geocentric model. Later, he served as his assistant.

To make his model compatible with experimental observations, Ptolemy was forced to introduce an imaginary point called "equant." He was unhappy about this imaginary point. Copernicus also disliked the concept of equant. From his teacher Maria Novara, he could think beyond an Earth-centered model; he knew of Philolaus' model where Earth rotates about a central fire (he mentioned about Philolaus in his book, *De Revolution*). It is uncertain whether he knew of Aristarchus of Samos, who was the first to put Sun at the central position. Copernicus wanted to check a

---

[7]A polyglot speaks five or more languages. Copernicus used to speak Latin, German, Polish, Greek and Italian.

[8]A polymath is a person of great and varied learning.

[9]A canon is a priest of the Cathedral Chapter appointed by the bishop to carry out various ecclesiastical functions.

Sun-centered model against experimental observation. From his chapel, he brought 800 building stones and a barrel of lime and built a small naked eye observatory and with few astronomical instruments, diligently looked into the sky. From his observations, he realized that the imaginary point equant could be dispensed with if instead of the Earth, Sun was the center of the Universe. Copernicus was aware that his Sun-centered or heliocentric model might not be tolerated by the Church and in 1514, anonymously published a handwritten booklet *Commentariolus* (usually translated as the *Little Commentary*) expounding his model of the Universe and distributed to a few friends. A schematic diagram of Copernicus' model of Universe is shown in Figure 3.4. The model was based upon seven "axioms":

 i. *There is no single center for all orbits in the Universe.*
 ii. *The Earth's center is not the center of the Universe, but only of the lunar orbit.*
 iii. *The center of the Universe is near the Sun.*
 iv. *The distance from the Earth to the Sun is imperceptible compared to the distance to the stars.*
 v. *The daily rotation of the Earth accounts for the apparent daily rotation of the stars, which themselves are immobile.*
 vi. *The apparent annual cycle of solar motion is caused by the Earth revolving round it once every year.*
 vii. *The apparent retrograde motion of the planets is caused by the motion of the Earth around the Sun, and from which one observes the planets.*

It is worthy of note that Copernicus did not put Sun at the center of the Universe. He was wary of the Church's reaction, even more, he himself was a deeply religious man. He put his observations as a model, rather than presenting it as truths of the nature. The fourth, fifth and sixth axioms exemplify the fundamental difference

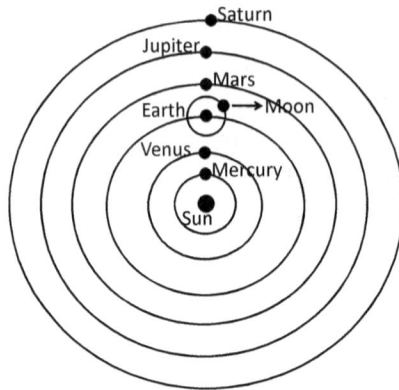

Figure 3.4. Copernicus' heliocentric model. As opposed to Ptolemy's model, Copernicus put the Sun at the center.

between heliocentrism and geocentrism. The Sun to Earth distance is imperceptible compared to Sun–stars distance. In one stroke, he enlarged the Universe much beyond than conceived by Aristotelians and countered the Aristotelian argument that for distant stars, parallax effect was not observed. In the fifth axiom, he introduced Earth's rotation causing day and night and in the sixth axiom, he made Earth rotate about the Sun. The retrograde motion of the planets, which led Ptolemy to devise the concept of epicycles, was then explained as due to the motion of the Earth around the Sun.

Indeed, in the Copernican model, the retrograde motion of a planet is easily understood. It is explained in Figure 3.5, where successive positions of the Earth and a planet are shown as they revolve around the Sun. If Earth completes the orbit faster than the planet then at some position Earth will overtake the planet and from the Earth the planet will appear to move backward, as in retrograde motion.

His second book on the subject, *De revolutionibus orbium coelestium* (*On the Revolution of the Celestial Spheres*), where the model was explained in detail, was published much later in 1543. He was on the verge of death. Indeed, Copernicus delayed the publication fearing possible furor over his model. He dedicated the

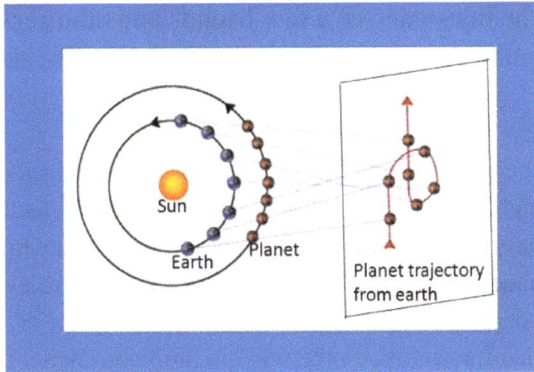

Figure 3.5. Retrograde motion of planet explained in heliocentric model.

book to Pope Paul III. The preface is rather revealing. It truly reflected the turmoil in his mind in going against the Church dictum. Below I reproduce a portion of the first paragraph of the preface,

"I can readily imagine, Holy Father, that as soon as some people hear that in this volume, which I have written about the revolutions of the spheres of the Universe, I ascribe certain motions to the terrestrial globe, they will shout that I must be immediately repudiated together with this belief For I am not so enamored of my own opinions that I disregard what others may think of them. I am aware that a philosopher's ideas are not subject to the judgment of ordinary persons, because it is his endeavor to seek the truth in all things, to the extent permitted to human reason by God. Yet I hold that completely erroneous views should be shunned. Those who know that the consensus of many centuries has sanctioned the conception that the Earth remains at rest in the middle of the heaven as its center would, I reflected, regard it as an insane pronouncement if I made the opposite assertion that the Earth moves. Therefore I debated with myself for a long time whether to publish the volume which I wrote to prove the Earth's motion or rather to follow the example of the Pythagoreans and certain others, who used to transmit philosophy's secrets only to kinsmen and friends..."

However, at the insistence of a few friends and admirers, Copenicus decided to publish the book. He ended the preface with the following:

> "Perhaps there will be babblers who claim to be judges of astronomy although completely ignorant of the subject and, badly distorting some passage of Scripture to their purpose, will dare to find fault with my undertaking and censure it. I disregard them even to the extent of despising their criticism as unfounded. For it is not unknown that Lactantius,[10] otherwise an illustrious writer but hardly an astronomer, speaks quite childishly about the Earth's shape, when he mocks those who declared that the Earth has the form of a globe."

Though Copernicus dedicated the book to Pope Paul III, possibly to obtain favorable reception from the Church, that however, was of no avail. It is difficult now to comprehend the enormity of Copernicus' discovery. In one stroke, he displaced Earth from its exalted position at the center of the Universe to an ordinary planet like Mercury, revolving around the Sun. We have been in tune, over a long time, with the Sun-centered model. Were we not, the idea would be staggering. The Earth's circumference at the equator is 40,075 km. Daily rotation of Earth implies that a point on the equator is moving at the enormous speed of 1670 km per hour. Even more staggering is the Earth's orbital speed. The average distance between the Earth and the Sun is 149,597,890 km. Then in one year Earth travels a distance of $2 \times \pi \times 149{,}597{,}890 = 939{,}950{,}470$ km. In a year, we have 8760 hours. Then the Earth's orbital speed is 107,300 km per hour. Imagine our solid Earth, with trees and houses, cities and countries, mountains and seas, envelope of air and all, rotating about itself at a speed of 1670 km/hour and rushing through the space at the speed of 107,300 km per hour. It is

---

[10] Lucius Lantantius (ca. 250–325) was advisor to the first Chrtistian Roman emperor Constantine I. He defended Christian beliefs and in his book, *Divinae Institutiones* (*Divine Institutes*) gave the first systematic Latin account of the Christian attitude towards life.

hard to believe even today, not to speak of centuries earlier. Indeed, Ptolemy argued that if Earth was subjected to this enormous motion, it would go asunder. Copernicus argued that Ptolemy's geocentric model envisaged even greater motion for the heaven and if heaven did not go asunder, there was no reason to believe Earth could not withstand much reduced motion.

While those who held on to the Aristotelian view opposed *De revolutionibus*, initially it met no resistance from the Catholic Church. Copernicus was a revered member of the Church (throughout his life, he held the canon's position). The book was patronized by Cardinal of Capua, Nicholas Schoeberg, who entreated its publication and also bore the cost of publication. The book was dedicated to the Pope. However, very soon, Catholic Church came under pressure from the Protestants. Martin Luther, the seminal figure in Protestant reformation, ridiculed Copernicus,

"This fool (Copernicus) wishes to reverse the entire science of astronomy; but sacred Scripture tells us that Joshua commanded the Sun to stand still, and not the Earth."

Indeed, the main argument against the Copernican view was that it was against the Christian belief and Scriptures. Following passages from Book of Joshua[11] (X-12-13) were often quoted to argue against the Copernican view,

12. Then Joshua spoke to the Lord in the day when the Lord delivered up the Amorites before the children of Israel, and he said in the sight of Israel:

   "Sun, stand still over Gibeon;
   And Moon, in the Valley of Aijalon."

---

[11] The Book of Joshua is the sixth book in the Hebrew Bible and the Christian Old Testament. The book tells of the story of Israelites' entry into the land of Canaan (present day Israel, Lebanon, Syria and Jordan), their conquest and division of the land under the leadership of Joshua, a biblical figure.

13. So the sun stood still,
    And the moon stopped,
    Till the people had revenge
    Upon their enemies.

    Is this not written in the Book of Jasher? So the Sun stood
    still in the midst of heaven, and did not hasten to go down for
    about a whole day.

Under pressure, in 1616, Catholic Church banned the book.
Incidentally, the ban continued for more than 200 years and was
finally lifted in 1835.

3.3 Oh no! Circle is not the right path

Even though the Copernican heliocentric model with the Sun at the
center was a great improvement over earlier models for the
Universe, still it was not a correct model. Ancient astronomers were
overwhelmed by Aristotle's notion of circular motion for the heavenly bodies. Like Ptolemy, Copernicus also could not go beyond the
circular motion for the planets and had to use epicycles (though
fewer than in Ptolemy's geocentric model), to correctly predict for
the planetary motion. Even then, a Copernican heliocentric model
could not predict the positions of the planets Venus and Saturn, correctly over a long time period. It was then a prevailing custom to
readjust frequently the positions of the planets. The correct model
was given much later by Johannes Kepler (December 27, 1571–
November 15, 1630). Kepler was a German mathematician and
astronomer. Born in a poor German protestant family, his love for
astronomy grew early when he (together with millions of Europeans)
observed the great comet[12] of 1577 and a lunar eclipse in 1580. The

---

[12] Comets are icy, small (40–100 km size) solar system bodies that when passing
close to the Sun, heat up and outgas. The dust and gases form a tail that stretches
away from the Sun for millions of kilometers. The original Greek word *kometes*
means long hair.

love generated from those fascinating sights continued throughout his lifetime. He was an exceptionally brilliant student and earned a scholarship to attend the University of Tuebingen. There he learned theology and philosophy. He also learned mathematics and astronomy. Teaching of the Copernican heliocentric model was then prohibited by the Church but his teacher, who was a believer of the heliocentric model, used to teach it secretly to brilliant students. From his teacher Kepler learned both the Ptolemaic and Copernican system of planetary motion. He intuitively sensed that the Copernican model of the Universe was correct and made it his mission to prove rigorously what Copernicus had only guessed.

On completion of education, Kepler became a teacher in a school at Graz, Austria. He was particularly intrigued by the number of planets. At that time only six planets were known: Mercury, Venus, Earth, Mars, Jupiter and Saturn. He wondered why there are only six planets, why not seventeen or even a hundred. As a teacher, Kepler had to teach various subjects. Once he was teaching a class on regular polygons. Regular polygons are closed planar figures where all the sides, as well as all the angles, are equal. A few examples of regular polygons are shown in Figure 3.6. Regular triangle (also called equilateral triangle) is the lowest ordered polygon that can be constructed. While teaching, Kepler realized that regular polygons bound one inscribed and one circumscribed circle at definite ratios.

As an example, in Figure 3.7, a regular triangle and the inscribed and circumscribed circles are shown. In a regular triangle, all the angles are equal and since the sum of the three angles is 180°, each

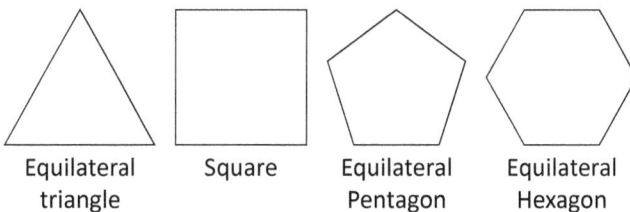

| Equilateral triangle | Square | Equilateral Pentagon | Equilateral Hexagon |

Figure 3.6. Examples of a few regular polygons.

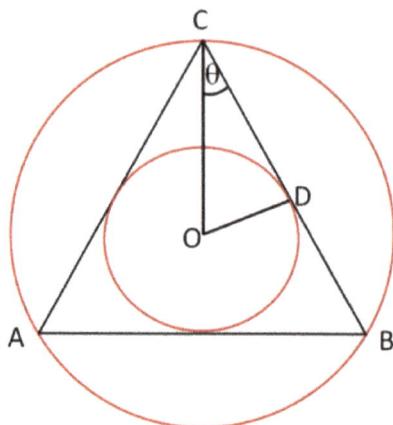

Figure 3.7.   Inscribed and circumscribed circles of a triangle.

of them corresponds to 60°. The radius OC of the circumscribed circle bisects the angle ∠ACB and the radius OD of the inscribed circle is perpendicular to the triangle side BC. The triangle OCD is then a right angled triangle. It is easy to calculate the ratio of the radius of the inscribed circle (OD) to circumscribed circle (OC),

$$OD/OC = \sin \theta = \sin 30° = 1/2.$$

Whatever be the size of the triangle, the ratio is a fixed number. Similarly, the ratio of the radii of the inscribed to circumscribed circles for a square is $1/\sqrt{2}$, again a fixed number. Indeed, one can prove a general result that for an N-sided regular polygon, the ratio of radii of inscribed to circumscribed circle is $\cos(\pi/N)$.

Kepler noted that for a triangle, the ratio 1/2 of the inscribed and circumscribed circle's radii is the same as the ratio of orbital radii of Saturn and Jupiter. Kepler did not believe that the coincidence of ratios was purely fortuitous, rather he reasoned that it might be the geometrical basis of the Universe. He thought that given Saturn and Jupiter were the first two planets (from outside) of the solar system, they orbited the inscribed and circumscribed circles of the simplest polygon, the triangle. He extended the reasoning to the ratio of the orbital radii of Jupiter and Mars. It should be in the same ratio as the radii of the circumscribed and inscribed circles of a square, the second simplest polygon, which is $1/\sqrt{2} = 0.876$. However, Kepler's reasoning failed, the ratio of the orbital radii of Jupiter and Mars turned out to be much smaller — approximately 0.05. Kepler tried several arrangements of regular polygons to fit the observed astronomical data without success. Then he had an inspiration. Polygons are planar figures, in two dimensions. The world, however, is three-dimensional. Why not try with three-dimensional analogs of regular polygons? Three-dimensional analogs of regular polygons are known from antiquity and are called Platonic solids because the Greek philosopher Plato wrote about them. A platonic solid is a polyhedron i.e. a solid with flat faces and that every face is a regular polygon of same size and shape. As shown in Figure 3.8, there are only five Platonic solids. It became interesting, for six planets, there will be five intervening spaces, to be occupied by five platonic solids!

Kepler found that each of the five Platonic solids could be uniquely inscribed and circumscribed by celestial spheres. Nesting

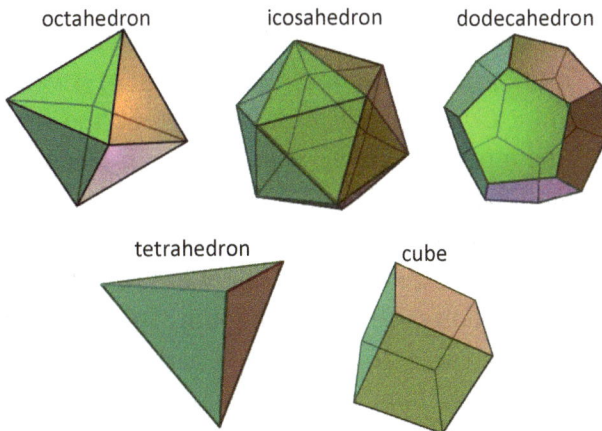

Figure 3.8. Five Platonic solids are shown.

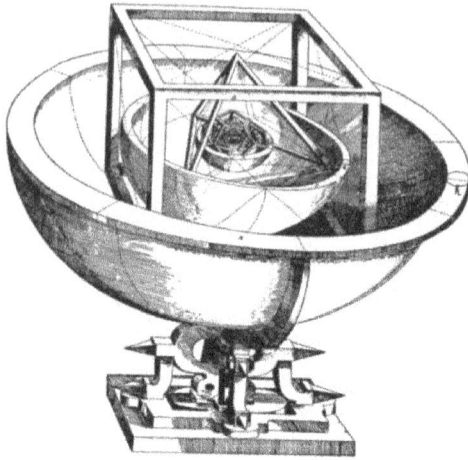

Figure 3.9.   Kepler's solar system based on Platonic solids.

these solids, each encased in a sphere, within one another would produce six layers, corresponding to the six known planets — Mercury, Venus, Earth, Mars, Jupiter, and Saturn. Assuming the planets encircle the Sun and by ordering the solids correctly — octahedron, icosahedron, dodecahedron, tetrahedron and cube — Kepler found that the spheres could be placed at intervals corresponding to the relative sizes of each planet's path. Figure 3.9 shows a model of Kepler's solar system based on Platonic solids. In 1596, Kepler published results of his investigations in a book, *Mysterium Cosmographicum*, or *Cosmographic Mystery*.

In *Mysterium Cosmographicum*, Kepler took the first tentative step towards the modern picture, where Sun — by its gravitation — controls the motions of the planets. In the first edition, Kepler attributed to the Sun a *motricem animam* ("moving soul"), which caused the motion of planets. Aristotle used the word to indicate "soul" or "life to animate" which enables a living thing to do what it properly does — a plant to grow, a dog to run and bark, a person to talk and think. In the same vein, Sun had a soul which controled the planets. In the second edition he came further closer to the modern picture, he supposed that some force — which, like light, is "corporeal" but "immaterial" — emanated from the Sun and drived the planets. While *Mysterium Cosmographicum* established Kepler as a respected astronomer, it was a false problem he was working on. Now we know there are more than six planets and no compelling geometrical reasons why the planets are at just the distances they are from the Sun. Kepler's major contribution to astronomy came a few years later. Kepler was a Lutheran Protestants. In 1600, for his religious believe Kepler was forced to leave Graz and moved to Prague. At Prague, he became an assistant to the then famous, Danish scientist and astronomer Tycho Brahe (December 14, 1546– October 24, 1601).

Son of a Danish nobleman, Brahe led a colorful life. When Brahe was two years old, he was stolen by his uncle, an act, surprisingly, not resented by his parents or he himself. He was interested in astronomy from his early age when he witnessed the solar eclipse in August 21, 1560. He was impressed by the sight but even more impressed by the fact that it was a predicted event. However, following the wish of his uncle, he studied law in the University of Copenhagen. When he was twenty, he lost his nose in a duel and for the rest of his life used a prosthetic nose, made of gold and silver. Purportedly, the cause of the duel was a heated discussion between Brahe and another Danish nobleman over some fine point of mathematics.

Brahe was not a believer of the heliocentric model (he could not feel Earth's motion!). He proposed a variant of the geocentric model where the Earth was fixed and Sun and Moon orbited the Earth while other planets orbited the Sun. He understood that his model could be established only if accurate measurements of planetary positions were obtained and devoted his life to that end. With the help of his patron, the Danish King Rudolf II, Brahe built the most accurate naked eye observatory ever. With the help of the huge observatory, he could measure planetary positions accurate to one arcminute[13] (an enviable task in pre-telescope days). Accurate clocks, invented by that time, also helped him. For the mathematical modeling of his model, Brahe appointed Kepler as his assistant. While the two had working relation, Brahe was distrustful of Kepler and closely guarded his measured data on planetary positions. Only after Brahe's death in October 1601 could Kepler access the large amount of data collected by Brahe (not purely by legal means).

After many trials and errors, Kepler understood the problem of the heliocentric model. Till then, astronomers were obsessed with Plato's idea of "perfect circular motion" — the heavenly bodies could only orbit in a circular path. Circle is the simplest closed curve. As shown in Figure 3.10, it has a center O, from where distance to any point on the curve is a constant. Thus for two points P and P' on the circles, OP = OP'. Ellipse is a generalization of the circle. As shown in Figure 3.10, it is also a closed curve but with two focal points, O and O' such that the line joining from one focal point to the curve and back to the second focal point is always a constant. Thus for two points P and P' on the ellipse, OP + PO' = OP' + P'O'.

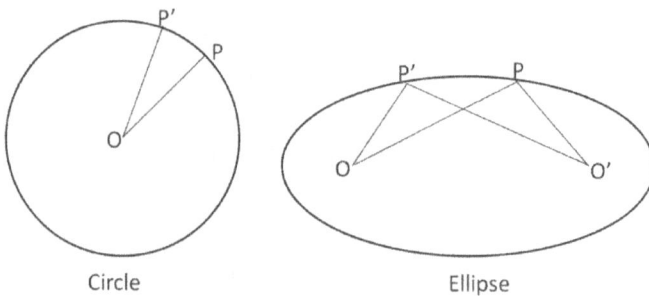

Figure 3.10.   Schematic diagram of a circle and an ellipse.

---

[13] Arcminute is a unit of angle measurement. One arcminute is equal to 1/60th of one degree.

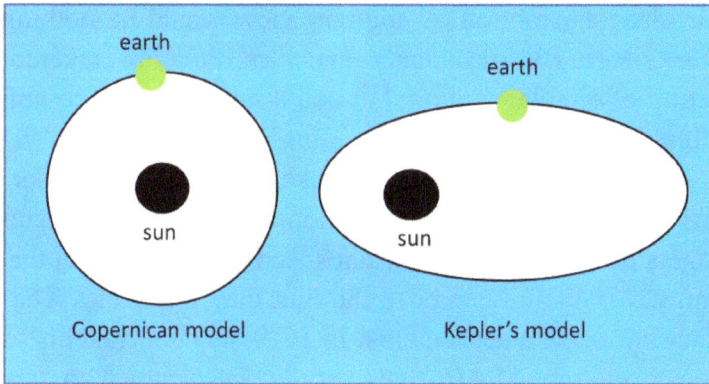

Figure 3.11. Schematic representation of a planetary orbit in Copernican and in Keplerian model.

Kepler noted that if the idea of the circular path could be dispensed with, most of the problems of the heliocentric model would be eliminated. He generalized the circular orbits to elliptical orbits and put Sun at one of the foci. In Figure 3.11, a schematic diagram for a planetary orbit in a Copernican and in a Keplerian model is shown. In 1609, he published his account of lifelong research, one of the greatest books in astronomy, the *Astronomia Nova* (*A New Astronomy*). He showed that the three models, the Ptolemaic model, the Copernican model and Tycho Brahe's model, are the same from observational point of view. In *Astronomia Nova*, Kepler also enunciated first two of his three planetary laws which are still valid,

1. *The orbits of the planets are ellipses, with the Sun at one focus of the ellipse.*
2. *The line joining the planet to the Sun sweeps out equal areas in equal times as the planet travels around the ellipse.*

It may not seem so now, but the first law was a giant leap in men's thinking process. For ages, men had followed Aristotle's dictum of circular planetary motion. A contrary view would be nothing less than a sacrilege. By introducing epicycles, Ptolemy tinkered with the dictum of circular motion, but could not come out completely. That Kepler, who said, "*Truth is daughter of time and I feel no shame in being her midwife*," could take that bold step speaks voluminously of his courage and conviction. The second law states that the planets do not move at constant speed. Far away from the Sun, they move slowly and nearer to the Sun they move fast. The third planetary law came a decade later. In 1619, he published his second work on cosmology, *The Harmony of the World*, which contains the third law,

3. *The ratio of the squares of the revolutionary periods for two planets is equal to the ratio of the cubes of their semimajor axes.*[14]

If P_1 and P_2 are period of revolution of two planets and R_1 and R_2 are their semimajor axis, Kepler's law can be written as,

$$\frac{P_1^2}{P_2^2} = \frac{R_1^3}{R_2^3}.$$

If rotation of a planet is measured in unit of Earth's rotation period (1 year) and semimajor axis is measured in unit of Earth's distance to Sun, called astronomical unit (AU), then, Kepler's 3rd law for the relation between the semimajor axis and the planetary period simplifies as,

$$R(\text{AU}) = P(\text{year})^{2/3}.$$

Consider the planet Mars. It orbits around the Sun in 686.97 days, or in 1.88 years. Using Kepler's 3rd law, semimajor axis of Mars

---

[14] Semimajor axis of an ellipse is half the sum of smallest and greatest distance from one of its focus.

can be easily calculated to be 1.52 AU, which indeed is the measured value.

3.4 Seeing the heaven through a spyglass

Astronomia Nova was published in 1609. However, the world at large stuck with the Ptolemaic model. The model, for all practical purpose served well. The world needed a genius like Galileo Galilei to (literally) show the wrongs, inadequacies of Aristotelian–Ptolemaic geocentric model. Galileo Galilei (February 15, 1564–January 8, 1642) is widely regarded as the father of modern science. One of the greatest scientists of all time, Galileo was an engineer par excellence and could be credited with the first gas thermometer, a military compass, hydrostatic balance for weighing, design of the first pendulum clock (Christiaan Huygens[15] built it in 1657), and the first compound microscope. Many of his inventions were to augment his income to support his family.

Born in Florence, Galileo had his early education in a monastery and intended to become a priest. But bowing to his father's interest, he joined the University of Pisa for a degree in medicine. There he found his real interest — mathematics and philosophy. Without completing the university education, he returned back to Florence and began to teach mathematics privately. He must have been a very successful teacher as later he obtained a public appointment as a tutor. In 1586, at the age of 22, he invented a hydrostatic balance to measure small objects and published the design in a brief treatise entitled *La Bilancetta*, or *"The Little Balance."* The "Eureka" story about Archimedes and the bath tub was well known

---

[15] Christiaan Huygens (April 14, 1629–July 8, 1695), a Dutch mathematician and physicist, was a prominent scientist of his time. He founded the wave theory of light, discovered the pendulum clock, the true shape of the rings of Saturn and made original contributions to the science of dynamics.

in Galileo's day as it is in ours. The story goes like this: King of Syracuse suspected his goldsmith of unfair practices and asked Archimedes to find out whether or not a particular crown is made of pure gold or mixed with silver, of course without destroying the crown. Archimedes pondered over the problem for days and nights but could not find a solution. Then one day he went to a public bath and as he slid into the tub full of water, large amount of water splashed over the brim; simultaneously, he felt himself to be lighter. The solution of the problem struck him like a lightning. He jumped out of the tub and naked, ran through into the streets of Syracuse, shouting "eureka, eureka," translated, "I found it, I found it." He had discovered the law of buoyancy — that an object immersed in liquid is acted upon by an upward thrust equal to the weight of the liquid displaced by it. He correctly reasoned that silver being lighter than gold, a crown mixed with silver would be bulkier to reach the same weight and would displace more water than its pure gold counterpart or a gold body with the same weight. In the story Archimedes then measured the amount of water displaced by the crown and by an equal weight of gold, and comparing the two found out whether the crown was of pure gold or not. Weighing precious metals in air and then in water was presumably a practice that was common among jewelers in Europe.

Galileo argued that Archimedes, who was well versed in the lever[16] action and who could make the bold statement:

"Give me but a firm spot on which to stand, and I shall move the Earth."

would not adopt such a crude method, as it would be rather inaccurate. Archimedes must have designed a balance based on

---

[16]A lever is a simple machine, consists of a stick or rod attached to a pivot or fulcrum. It helps to lift weight with less effort. The children's see-saw is an example of lever.

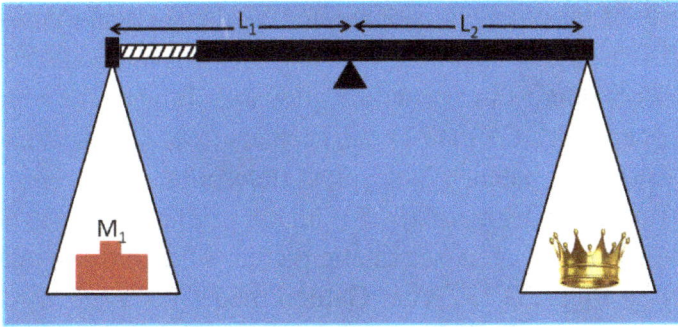

Figure 3.12. Schematic picture of Galileo's hydrostatic balance.

lever principle and he described one such balance. A schematic picture of the balance is shown in Figure 3.12. One arm of the balance, by screw and wire arrangement, can be shortened or lengthened. Say, the crown is balanced by a weight M_1, then from the principle of lever action,

$$\text{Weight of crown} \times L_2 = M_1 \times L_1.$$

When suspended in water, the crown will weigh less and the screw–wire arrangement can be used to shorten the counter arm L_1 to balance the crown again and very accurate measure of the new weight can be obtained.

The clever design of Galileo was noted by the scholarly world and in 1589 he was appointed to the chair of mathematics at Pisa. At Pisa, Galileo conducted the famous experiment with falling objects. He dropped two balls of different masses from the Tower of Pisa and demonstrated that time of descent was the same for the two masses. At that time Aristotelian view prevailed that a heavy object would fall faster than a lighter one. Direct demonstration contradicting Aristotelian view made Galileo famous. In 1592, he moved to the University of Padua, teaching geometry, mechanics, and astronomy. Following his invention of telescope, Galileo's fame grew and in 1610, he was appointed for lifetime, "Chief Mathematician of the

University of Pisa and Philosopher and Mathematician to the Grand Duke of Tuscany."[17]

Galileo's interest in astronomy grew late. In 1609, he heard that Hans Lippershey, a German-Dutch spectacle-watch maker, had filed an application for patent of a spyglass (telescope) *"for seeing things far away as if they were nearby."* With his engineering skill, Galileo was able to make one. A schematic picture of a Galilean telescope is shown in Figure 3.13. What Galileo made was a refractive telescope, i.e. he used the refractive property of light. Light rays from a distant object are refracted by the primary lens with focal length[18] f_0 and seen through a secondary lens, called eyepiece, with focal length f_1. Magnification of the object can be calculated as the ratio of the focal lengths, $M = f_0/f_1$.

It was fortunate for the mankind that Galileo turned his telescope to the sky. What he observed excited him so much that he

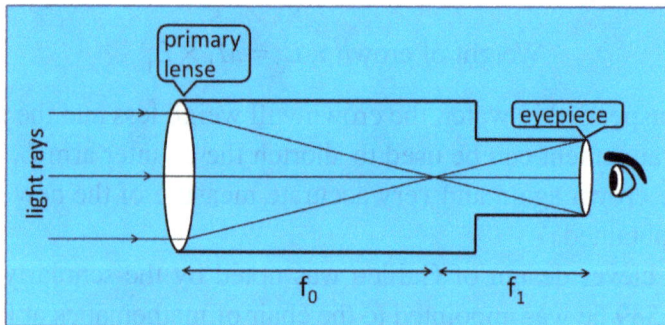

Figure 3.13.   Schematic picture of a Galilean telescope.

---

[17] Tuscany is a region in central Italy with an area of about 23,000 square kilometers and a population of about four million inhabitants (2013). The regional capital is Florence. Cosimo II de' Medici was the Duke of Tuscany at the time of Galileo.

[18] Focal length of a lens is the distance over which initially collimated rays are brought to a focus, i.e. parallel rays are focused to a point.

immediately published results of his observations in a brief treatise entitled *Sidereus Nuncius* (*Starry Messenger*). He wrote,

"Great indeed are the things which in this brief treatise I propose for observation and consideration by all students of nature. I say great, because of the excellence of the subject itself, the entirely unexpected and novel character of these things, and finally because of the instrument by means of which they have been revealed to our senses... It is a very beautiful thing, and most gratifying to the sight, to behold the body of the Moon, distant from us almost sixty earthly radii as if it were no farther away than two such measures — so that its diameter appears almost thirty times larger, its surface nearly nine hundred times, and its volume twenty-seven thousand times as large as when viewed with the naked eye. In this way one may learn with all the certainty of sense evidence that the Moon is not robed in a smooth and polished surface but is in fact rough and uneven, covered everywhere, just like the Earth's surface, with huge prominences, deep valleys, and chasms."

With the help of his spyglass, he observed uneven surface of the Moon, craters and mountains, phases of Venus and four moons of Jupiter, all of which were in direct contradiction to Aristotelian views or to Ptolemy's geocentric model. On observing the Jupiter's moon he became an open supporter of Copernican heliocentric model and wrote,

"Here we have a fine and elegant argument for quieting the doubts of those who, while accepting with tranquil mind the revolutions of the planets about the Sun in the Copernican system, are mightily disturbed to have the Moon alone revolve about the Earth and accompany it in an annual rotation about the Sun. Some have believed that this structure of the Universe should be rejected as impossible. But now we have not just one planet rotating about another while both run through a great orbit around the Sun; our own eyes show us four stars which wander around Jupiter as does the Moon around the Earth, while all together trace out a grand revolution about the Sun in the space of twelve years."

In 1613, he openly supported the Copernican view of solar model and fell into conflict with the Roman Catholic Church. In 1616, the Church formally banned Galileo's writings, and warned him not to *"hold or defend his doctrines."* After eight long years, Galileo appealed to Pope Urban VIII, to revoke the ban. Not revoking the ban, the Pope allowed Galileo to write on both sides of the issue, without making any conclusion. Galileo took the opportunity to write his famous book, *Dialogue Concerning the Two Chief World Systems.* The book was published in 1632. It was written as a dialogue between Salviati, a philosopher believing in the Copernican system; Simplicio, a philosopher; believing in the Aristotelian and Ptolemaic system; and Sagredo, an intelligent layman. The book angered the Pope. The Pope was even more angry as Galileo put Pope's arguments for the Ptolemaic system in the mouth of Simplicio. *"Simplicio"* in Italian had the connotation of simpleton or fool. The book was banned and an inquisition was ordered. Galileo was held guilty of "heresy" (provocative belief or theory that is strongly at variance with established beliefs or customs). He was offered either to denounce the heliocentric model, publicly, or be branded as a heretic (which eventually would result in a painful death). Seventy years old, nearly blind, Galileo chose to renounce the Copernican belief: *"I held, as I still hold, as most true and indisputable, the opinion of Ptolemy, that is to say, the stability of the Earth and the motion of the Sun."* There is an unconfirmed story that after publicly renouncing the Copernican model, Galileo muttered, *"Yet it moves."* Galileo spent the remaining years under house arrest. While under house arrest, Galileo wrote *Two New Sciences,* a summary of his life's work on the science of motion and strength of materials, which he himself considered, *"superior to everything else of mine hitherto published."* On January 8, 1642, Galileo died after a prolonged illness.

3.5 Death at the altar of science

Account of historical development about cosmological models will be incomplete unless we mention Giordano Bruno (1548–1600), the free thinker who gave his life defending the heliocentric model.

Giordano Bruno was an Italian scientist and above all a free thinker. Son of a soldier, Bruno studied at a Naples monastery and at the age of 24, became an ordained priest. During his time in Naples, he became known for his skill with the art of memory and once traveled to Rome to demonstrate his mnemonic system to the Pope. While distinguished for outstanding ability, Bruno's taste for free thinking and forbidden books caused him difficulties. He fell into disgrace for studying the works of Erasmus[19] and left Naples in 1576. From 1576 to 1592, for sixteen long years he wandered into difference places of Europe: Austria, Switzerland, France, Germany and England. However, he could not settle anywhere. Wherever he went, he got involved into controversy, quarreled with his patrons and left for a newer place. While wandering, he wrote profusely: *The Shadow of Ideas* (1582), *The Art of Memory* (1582), *Incantation of Circe* (1582), *The Ash Wednesday Supper* (1584), *The Cabala of Pegasus* (1584), *On Cause, Principle and Unity* (1584), *The Expulsion of the Triumphant Beast* (1584), *On the Infinite, the Universe and Worlds* (1584), to name a few. He openly

---

[19] Desiderius Erasmus was a Dutch Renaissance humanist, Catholic priest, social critic, teacher, and theologian. He was critical of abuses of the Catholic Church and called for reform. His Latin and Greek editions of New Testament were influential in the Protestant Reformation.

advocated for the Copernican heliocentric model and unwittingly went beyond him. He said, "Sun" is just another star. He argued that since God is infinite, his creation Universe is also infinite and indeterminable. In 1584, Bruno published his account of cosmology, *On the Infinite, the Universe and Worlds* where an infinite Universe is filled with a substance — "a pure air" or aether — that offered no resistance to heavenly objects. He also declined to accept that Earth and Sun are special; to him they are one of many planets and stars of the Universe.

In March 1592, Bruno returned to Venice at the invitation of Giovanni Mocenigo, to teach him the art of memory. The Mocenigo family was one of the influential families in Venice. For about two months Bruno taught Giovanni, and then desired to leave his position. Unsatisfied with Bruno, Giovanni accused Bruno of "heresy and blasphemy." For about a year, he was tried at Venice and then extradited to Rome. For seven years, Roman Inquisition tried him for the following charges:

   (i) Holding opinions contrary to the Catholic faith and speaking against it and its ministers;

  (ii) Holding opinions contrary to the Catholic faith about the Trinity, divinity of Christ, and Incarnation;

 (iii) Holding opinions contrary to the Catholic faith pertaining to Jesus as Christ;

 (iv) Holding opinions contrary to the Catholic faith regarding the virginity of Mary, mother of Jesus;

  (v) Holding opinions contrary to the Catholic faith about both Transubstantiation and Mass;

 (vi) Claiming the existence of a plurality of worlds and their eternity;

(vii) Believing in metempsychosis and in the transmigration of the human soul into brutes;

(viii) Dealing in magics and divination.

While Bruno defended himself magnificently, it was of no avail. Inquisition found him guilty of heresy. He was given the opportunity to publicly denounce the Copernican model, but he declined. Inquisitors indicted him *to be an impenitent, pertinacious, and obstinate heretic.* Upon his conviction, Bruno reportedly responded:

"Perchance you who pronounce my sentence are in greater fear than I who receive it."

Roman court deliberated over his punishment and ultimately sentenced him to death: to be burned at the stake.[20] On February 17, 1600, the sentence was carried out. Life of a free thinker was extinguished.

One may wonder, why Kepler who was a declared Copernican was not condemned by the Church as was Galileo or Giordano Bruno? The reason is obscure. At that time, Europe was going through the religious conflict between Protestants and Catholics. Kepler was a Protestant, while Galileo and Bruno were Catholic. This may be the reason why Kepler was not subjected to the severe persecution by the Catholic Inquisition. Incidentally, Kepler's mother was repeatedly charged with witchcraft, and Kepler spent much of his time trying to free her of those charges.

3.6 How fast the light moves?

Galileo, through his spyglass observed four moons of the Jupiter. He suggested that periodic eclipse of Jupiter's moons could serve as a universal clock and greatly help navigation. While his suggestion did not fructify, Jupiter's moon gave the astronomers the first tool to measure the speed of light. Light plays a central role in our life.

---

[20] Burned at the stake was a form of execution practiced in the medieval Europe. The victim was bound with rope and chain to a wooden stake (long pole) and slowly burned to death.

Unarguably, life on Earth would not be possible without the solar light. It is also a universal messenger; it brings information from our surroundings and also from distant stars and galaxies. It tells us about the distant past. It also played an important role in the development of two fundamental theories of the physical world — the quantum mechanics, the mechanics obeyed by atomic and subatomic particles; and relativistic mechanics, the mechanics for bodies moving at a speed comparable to speed of light. Albert Einstein developed his theory of relativity from the postulate that nothing can move faster than light.

The speed of light intrigued early philosophers and scientists. In everyday experiences, light moves instantaneously. Aristotle believed that the speed of light is infinite. Though there were detractors (for example the Greek philosopher Empedocles believed in finite light speed), Aristotelian belief prevailed till Galileo's time. The most common argument for the infinite speed was that when a canon is fired from a great distance, one immediately see the light, but sound is heard fairly later. Galileo did not agree and said that the above example only proved that light traveled faster than sound and nothing more. True to his scientific spirit, he devised an experiment to measure the speed of light. Two men with covered lantern were put on two hilltops one mile apart. One man opened the cover of lantern, on seeing the light, the other immediately uncovered his. Taking into account the delay for human reaction, Galileo could only conclude that light traveled at least ten times faster than sound. Now we understand the reason. The speed of light is great: 186,000 miles per second. Two miles distance will be covered in 2/186,000 seconds, too small to be detected by clocks then available. The Speed of sound is approximately 340 meters or 0.21 miles per second. Enormity of the speed of light could not be comprehended from Galileo's conclusion that light travels at least ten times faster than sound.

Comprehension about the enormity of the speed of light first came from a measurement by the Danish astronomer Ole Roemer

(September 25, 1644–September 19, 1710). Roemer had his education at Copenhagen. Later he moved to Paris, where King Louis XIV employed him as the tutor to Dauphin (heir apparent to the throne of France). He also worked as an assistant to Giovanni Domenico Cassini, the famous astronomer who discovered the rings of Saturn. The innermost moon of Jupiter is named Io and Cassini was studying its eclipse. He observed some discrepancies in Io's eclipse. Sometimes the eclipse occurred earlier than the predicted time and sometimes later. As Cassini's assistant, Roemer observed hundreds of eclipses of Io. He noted that the time between successive eclipses of Io was shorter when the Earth was close to Jupiter and longer when further away from it. He correctly reasoned that the discrepancy in Io's eclipse period was due to the finite speed of light. The basis of his arguments is depicted in Figure 3.14.

Say at Earth we are measuring Io's period of revolution by observing the eclipse of Io. The period of revolution can be found from the time difference between two consecutive emergences of Io from the shadow of Jupiter or two consecutive immersions into the shadow. Say we make two measurements of Io's orbital period six months apart. Roemer observed that there was a difference of 11 minutes between the two measurements. Now Earth's position

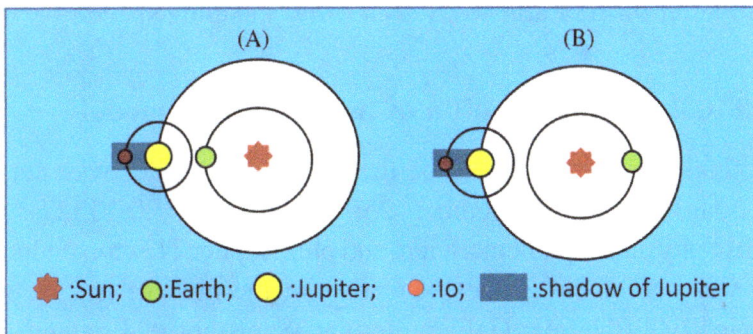

Figure 3.14. Schematic diagram of Roemer's determination of the velocity of light from eclipse of Io. Io's orbital period measured from two positions of Earth (A) and (B) shows a difference.

has nothing to do with Io's orbital period. Roemer correctly guessed that the difference is due to the extra distance, equal to the diameter of Earth's orbit around the Sun, light had to travel. From the known Earth's orbital diameter one can then find the speed of light. In reality, things are more complicated. Io's orbital period is approximately 1.77 days and during this time, Earth moves a finite distance. Roemer did not actually calculate the speed of light. Rather he sent his measurements to the physicist, Christiaan Huygens who calculated the speed of light as 214,390,000 meters per second, approximately 30% less that the currently accepted value, 299,792,458 meters per second. While the measurement was not very accurate, still, mankind obtained the first hint of the enormity of speed of light.

More accurate measurement of speed of light had to wait a century. In 1849, French physicist Hyppolyte Fizeau (September 23, 1819–September 18, 1896) measured speed of light as 308,448,000 meters per second, accurate to within 5%. The most accurate measurement of speed of light was due to Albert Abraham Michelson, the first American physicist to receive Nobel Prize in physics. He measured speed of light to be c = 298,666,666 meters per second. The third English letter 'c' is now the universal symbol for the speed of light. If you ask why c, the probable answer is that it is the initial letter of the Latin word "*celeritas*" meaning speed.

3.7 Final nail in the coffin of the geocentric model

The final nail in the coffin of the geocentric model was put by Sir Isaac Newton (December 25, 1642–March 20, 1727), the English physicist, mathematician, and philosopher. Newton undoubtedly was one of the greatest scientists of the modern era. He was born in 1642 (the year Galileo died) at Woolsthorpe, England. His father, a wealthy farmer, died three months before his birth. His mother remarried when he was three years old, and Newton was raised by his grandparents. He had early education at Kings School,

Grantham, where he did not excel. He was a premature baby and suffered from ill health during his childhood. Due to poor health, he generally avoided usual school games and spent time in developing sundial, windmill (which actually milled wheat) etc. His study was interrupted at the age of 12, when his mother, second time widowed, tried to make a farmer out of him. However, Newton, who in future would be a giant in science, was a complete failure as a farmer. He was sent back to school. During his schooling, Newton did not show any indication of scholarship. On completion of the school education, in 1661 Newton joined Trinity College, Cambridge, as a *sizer* (sizers were allowed free education, in consideration of performing certain, at one time menial, duties) and received his bachelor's degree in 1665. While at Cambridge, Newton started to keep record of wide-ranging topics under the heading *Certain Philosophical Questions*. Later, it became famous as Newton's Note Book. The entries are interesting. For example in one of the entries, he noted an easy way to enlarge or compress a picture:

"Make a square, and then divide it into many equal parts with the compasses, and draw one side to another lines with a ruler and a pencil, so that the picture be divided into equal squares, and so make squares on a fair paper as little or big as you will, but let there be so many, as there is in the picture, then observing the order of the squares draw the picture over with a pencil passing from square to square."

At Cambridge also, Newton had rather an indistinct career as a student. The BA degree was without honors or distinction, but sufficiently good to earn a scholarship to continue his study. In 1665–66, London witnessed the last of plague epidemics,[21] which

---

[21] London witnessed a series of bubonic plague in 17th century. There had been 30,000 deaths due to the plague in 1603; 35,000 in 1625; 10,000 in 1636 as well as smaller numbers in other years.

killed 100,000 of London's rapidly expanding population of about 460,000. Cambridge was closed as a precaution against the Great Plague and Newton returned to his home at Woolsthorpe. Two years of private study at home saw the development of his theories on calculus (Newton called it Fluxion), the law of gravitation, three laws of motion, theory of light and colors etc. In April 1667, when the threat of plague subsided, Newton returned to Cambridge. There he befriended the mathematician, Isaac Barrow, the first Lucasian Professor.[22] Barrow was greatly impressed by Newton's mathematical talent. On his urging, Newton published his treatise on infinite series, *On Analysis of Infinite Series*, that too only after a fellow mathematician published some results which Newton obtained earlier during his home study. That was a trait in Newton's character. He was reluctant to publish the results of his investigations. For example, his magnum opus, *Philosophiæ Naturalis Principia Mathematica*, often referred to as *Principia*, was published only in 1687, that too at the insistence of Edmund Halley, who also bore the cost of publication. Most of the contents of *Principia* had been worked out during his home study in 1665–1667. In 1669 Barrow became Master of Trinity and he recommended Newton as his successor. So at the young age of 26, after receiving his MA degree, Newton became the Lucasian Professor at Cambridge.

Our modern understanding of light and color is due to Newton. Before him, it was believed that light acquired color only on passing through a medium. Robert Hooke (1635–1703), English scientist and philosopher, an important figure at Newton's time, held that view. From a series of experiments in 1662–1667, Newton knew that white light was a combination of seven different colors found

---

[22] Reverend Henry Lucas, a member of British parliament, bequeathed an income of 100 pound per year to Cambridge University to establish a chair in mathematics. The chair is known as Lucasian Professor.

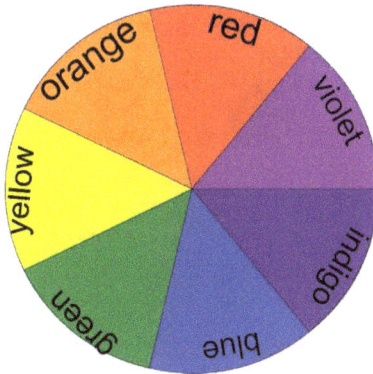

Figure 3.15. Newton's disc.

in a rainbow. In a simple experiment, he demonstrated his idea. White light was passed through a prism to be decomposed into seven colors: red, orange, yellow, green, blue, indigo and violet. Then using a second prism he merged the seven colored light again to white light. He also made a simple device to demonstrate the seven colors of light. As shown in Figure 3.15, he painted a disc in seven colors found in rainbow: violet, indigo, blue, green, yellow, orange and red (VIBGYOR). If the disc is rotated, the colors fade to white.

In 1668 Newton invented the reflecting telescope which changed the world of astronomy. It enhanced Newton's reputation among the scholarly community. Galilean telescopes used the refractive property of light. In a refractive telescope, light from a celestial body is refracted by a glass lens and forms an image at the eye piece. Images formed in Galilean telescopes were never very sharp. Newton, from his experiments with prisms knew that on passing through glass, different colors are bend by different amounts, most for the blue light and least for the red light. Thus image of a point source of white light formed by refraction through a lens will never be a point, rather it will be spread out, diffused. This effect, failure of a lens to focus all the colors to the same convergence point, is

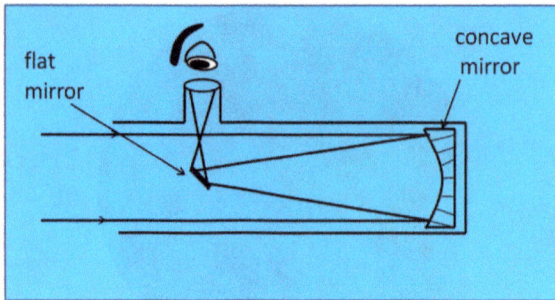

Figure 3.16.   Schematic diagram of a reflecting telescope.

called chromatic aberration. Chromatic aberration however is absent if the image is formed using reflective property of light. In his telescope, instead of a lens, Newton used a concave mirror wherein, the aberration problem was averted. A schematic diagram of Newtonian telescope is shown in Figure 3.16. Light rays from the object are reflected by the primary concave mirror to a secondary flat mirror to be directed to the eye piece. Much clearer and brighter images of heavenly bodies can be seen. Newton's reputation among the scholarly world grew when in 1671, at the urging of Barrow, he demonstrated his reflecting telescope (made in 1668) to the members of the Royal Society, London. Subsequently, he became a Fellow of the society. Newton continued at Cambridge till 1696, when he moved to London to become "Warden of Mint" and later in 1700 "Master of Mint."

Newton's achievements were numerous. He formulated the three laws of motion (known as Newton's laws of motion), inverse square law of gravity (showed that the force that make the planets go round the Sun and the force that pulls an apple towards Earth are the same), spectrum of light (visible light is made up of colors), reflecting telescope, method of calculus, binomial theorem and many more. Apart from *Principia*, in 1704, he published his second major work, *Optiks*, documenting his lifelong research on optics.

In *Principia*, Newton formulated three axioms, which are known as Newton's laws of the motion and completely characterize motion of classical bodies. For completeness, the laws (as enunciated by Newton in *Principia*) are mentioned here:

Law I: Everybody perseveres in its state of rest or of uniform motion in a right line, unless it is compelled to change that state by forces impressed thereon.

The first law is just a statement of inertia. A body likes to continue in its present state, unless acted upon by external force. This law may appear to contradict everyday experience; for example, if a ball is rolled on the floor, it eventually stops. The key point is the impressed force. When we roll the ball, it is acted on by friction force of the floor, friction force of the air etc. If they are removed, the ball will continue its motion. Newton's first law also has a serious implication about a fundamental symmetry of our Universe. *A state of motion in straight line is indistinguishable from the state of rest.* Say, you are in an immobile railway carriage, and playing with a ball, throwing it up and catching again. Let the carriage move with uniform speed. You can continue to play with the ball without any extra effort for the motion of the carriage. Observing you, a co-passenger cannot tell whether the carriage is mobile or immobile.

Law II: The alteration of motion is ever proportional to the motive force impressed; and is made in the direction of the right line in which that force is impressed.

The second law is the most important law in mechanics. It defines one of the most important and fundamental concepts in science: mass. Newton used the word "mass" as a synonym for "quantity of matter." These days mass is defined as a "measure of the inertia of a body." The more massive an object the more difficult it is to change its state of motion. The law states that the net force on a body is equal to the mass of the body times the change in motion or acceleration.

> Law III: To every action there is always opposed an equal reaction: or the mutual actions of two bodies upon each other are always equal, and directed to contrary parts.

The third law appears to be trivial but it is subtle. When we sit on a chair, we are exerting a force on the chair and simultaneously, the chair is exerting a force on us exactly equal and in the opposite direction. How a bird flies? It exerts a force on the air by flapping its wings and the air exerts a force on the bird and let it fly.

In the next step, Newton connected his laws of motion to the planetary motions. He understood that circular or elliptical motion of planets indicated departures from the inertial or straight line paths and some unbalanced force was acting on the celestial bodies to cause this departure. For years he contemplated on that force. As the legend goes, one day, while sitting under an apple tree, an apple fell on his head and the solution struck him like a lightning. Newton asked himself, why did the apple fall? It fell because it was attracted by the Earth. The apple was also attracting the Earth, but its effect on Earth was imperceptible due to its huge mass. He contemplated a similar force of attraction between the heavenly bodies.

Now how do we get the closed orbits of motion? Newton performed a thought experiment. Suppose a cannonball is fired horizontally from a high mountain top. If the air resistance were neglected, in absence of the gravitational attraction of the Earth, the cannonball would continue to travel in a straight line. With gravitational attraction, the Earth will attract the cannonball (as an apple is attracted by the Earth) and instead of continuing in its initial straight line path the cannonball will fall below the straight line path and eventually fall to Earth (see the path B in Figure 3.17). If the cannonball is fired with higher energy (more gun powder), it will still fall but after traversing a greater distance (path C). Now, suppose

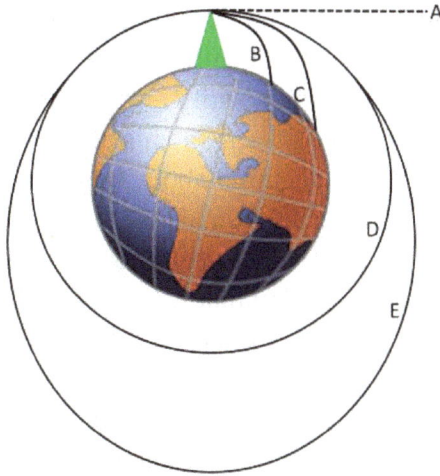

Figure 3.17. Newton's thought experiment with cannonball. A cannonball fired from a mountain top will continue in inertial or straight line path (depicted by the dashed line) in absence of gravity. In presence of gravity, the path will deviate from the straight line path.

the cannonball is fired still with higher energy such that trajectory of the cannonball matches the curvature of the Earth? The cannonball would fall towards the Earth without ever colliding into it and subsequently become a satellite orbiting in circular motion (as in path D). If fired with still higher energy, the ball will orbit the Earth in an elliptical path E.

Newton then formulated his famous universal law of gravitation,

> Any two objects exert a gravitational force of attraction on each other. The magnitude of the force is proportional to the product of the gravitational masses of the objects, and inversely proportional to the square of the distance between them. The direction of the force is along the line joining the objects.

Now, Newton had all the mathematical tools necessary for calculating the planetary orbits. He derived Kepler's laws of planetary motion from his theory of gravity, calculated the trajectories of comets, gave explanation of the tides, described the precession of the equinoxes, and many other phenomena, removing any doubt about the validity of the heliocentric or the Sun-centered model. However, as it was customary with Newton, he did not make his results public. As the story goes, British scientist Edmond Halley[23] went to visit Newton and enquired about the orbit of a planet under the inverse square law. Newton immediately answered that it would be an elliptical orbit. However, he could not provide the proof, as he had lost it and promised Halley that he would redo the calculations and send him the proof. Indeed, within a short period, Newton did those calculations again and dispatched to Halley the mathematical proof that under inverse square law, planetary orbits are ellipses.

While during the 18th century, Ptolemy's geocentric model was abandoned in favor of the heliocentric model, one question remained: despite so many interactions among the Sun and planets and satellites, how is it that the total system remains stable? Why is it not collapsing? Newton himself believed that some "divine" intervention was needed to obtain stability. In his letter to Richard Bentley,[24] he wrote,

> "... that the diurnal rotation of the planets could not be derived from gravity but required a divine power to impress them. And though gravity might give the planets a motion of descent towards the Sun either directly or with some little obliquity, yet the transverse motions by which they revolve in their several orbits required the divine arm to impress them ..."

It was left to Laplace, who, a century later solved the issue of stability of the solar system and showed that it did not require any

---

[23] British astronomer and mathematician who was the first to compute the orbit of a comet, later named after him, the Halley's Comet.

[24] Richard Bentley (1662–1742), Master of Trinity College, Cambridge, was an English scholar and theologian. He is called "founder of historical philology." He wanted to establish the existence of an intelligent creator for the Universe and corresponded with Newton on the subject.

divine intervention. Pierre-Simon marquis de Laplace (March 23, 1749–March 5, 1827) was a French mathematician and astronomer. His father was keen to see his son an ordained priest and got him admitted to Caen University to study theology. At Caen, Laplace discovered his mathematical ability and his love for the subject. He decided to have a career in mathematics. At the age of 19, without any degree, but armed with a recommendation letter from Le Canu, his teacher at Caen (who realized his great mathematical potential), he went to Paris to meet the famous French mathematician, physicist Professor Jean Le Rond d'Alembert (November 17, 1717–October 29, 1783). An illegitimate son of a French socialite, d'Alembert was left at the door of the Parisian church of Saint-Jean-le-Rond, from which his Christian name was derived. Though unacknowledged, his father left him considerable annuity and d'Alembert had a rather decent education. He studied law and became an advocate though never practiced. He studied medicine for a year and finally devoted himself to mathematics — *"the only occupation,"* he said later, *"which really interested me."* Even though most widely known for formulating the d'Alembert's principle of classical mechanics, he contributed immensely in integral calculus, mechanics, and partial differential equations. d'Alembert disliked students who came with recommendations. When Laplace sent him Canu's recommendation letter and asked for an audience, d'Alembert did not entertain him. Laplace then prepared a mathematical note and sent it to d'Alembert. d'Alembert immediately recognized his talent. In reply, he invited Laplace to meet him and wrote,

> "Sir, you see that I paid little enough attention to your recommendations; you don't need any. You have introduced yourself better."

Laplace became a protégé of d'Alembert who found a position for him at the École Militaire, to teach mathematics to the military cadets. Incidentally, there he examined and passed a 16-year-old cadet, Napoleon Bonaparte, who later became the Emperor of France. At Paris, Laplace worked hard and produced a steady stream of remarkable mathematical papers to establish himself as a gifted mathematician. His achievements are numerous and it is not unjustified that sometimes he was called Newton of France. Students of mathematics and physics are well acquainted with the Laplace transform. Laplacian differential operator, widely used in mathematics, is also named after him. He pioneered the probability theory. He also developed the nebular hypothesis of the origin of the solar system and was one of the first scientists to postulate the existence of black holes and the notion of gravitational collapse. In 1799, when Napoleon became Emperor of France, he appointed Laplace to the post of Minister of the Interior. The appointment, however, lasted only six weeks. Napoleon reportedly said:

"Geometrician of the first rank, Laplace was not long in showing himself a worse than average administrator; from his first actions in office we recognized our mistake. Laplace did not consider any question from the right angle: he sought subtleties everywhere, conceived only problems, and finally carried the spirit of 'infinitesimals' into the administration."

Among all his works, Laplace could be remembered only for his monumental work on celestial mechanics. Between 1799 and 1825 Laplace published his greatest work, the five volumes of *Treatise on Celestial Mechanics*. Even though Newton invented calculus, in *Principia* Newton did not use calculus, rather gave his derivations geometrically. Laplace's *Treatise on Celestial Mechanics* not only expressed Newton's *Principia* in modern language, in the language of the differential calculus, but it also completed parts which Newton had been unable to fill in the details. In *Celestial Mechanics*, Laplace borrowed heavily from his predecessors, Lagrange and Legendre (and most of the time without proper acknowledgement).

He developed the concept of a "potential" and he described it by an equation, now called Laplace equation. Laplace equation greatly simplified the study of the solar system, which is a many-body problem. He showed that the perturbations of the planetary orbits caused by the interactions of planetary gravitation are in fact periodic and that the solar system is, therefore, stable, requiring no divine intervention. His following remark on Arabic-Hindu numerals is fascinating.

> "It is India that gave us the ingenious method of expressing all numbers by means of ten symbols, each symbol receiving a value of position as well as an absolute value; a profound and important idea which appears so simple to us now that we ignore its true merit. But its very simplicity and the great ease which it has lent to computations put our arithmetic in the first rank of useful inventions; and we shall appreciate the grandeur of the achievement the more when we remember that it escaped the genius of Archimedes and Apollonius,[25] two of the greatest men produced by antiquity."

While Laplace solved the problem of stability of solar system, one question remained. How the solar system came to be? Is there an intelligent creator? Laplace can also be credited with the first credible theory of formation of solar system, the so-called "solar nebula" theory, very close to the presently accepted model of solar system formation. However, before we learn about the nebular theory, let us see in detail our solar system.

---

[25] Apollonius of Perga (ca. 240–190 BCE) was a Greek mathematician and geometer, known for his works on conic sections. The modern names ellipse, parabola, hyperbola are due to him. He greatly influenced Ptolemy, Copernicus, Newton and others.

Chapter 4

Modern Cosmology

> The phenomena of nature, especially those that fall under the inspection of the astronomer, are to be viewed, not only with the usual attention to facts as they occur, but with the eye of reason and experience.
>
> William Herschel

4.1 Solar system

If, in Plato's Academy, the question was asked: "How many planets are there?" students would have answered "Five." Greek astronomers, with their naked eyes, could find only five planets: Mercury, Venus, Mars, Jupiter, and Saturn. Moon of the Earth was the only moon or satellite known to them. They did not know or thought that Earth was also a planet. Number of planets remained fixed at five for long; even in 1632, Catholic Church forced Galileo to withdraw his support for the Copernican model which viewed Earth as just another planet. Towards the end of 17th century, when Newton conclusively proved the Keplerian model of Universe, Earth permanently entered into the list of planets. Even in Newton's time, when men looked through a telescope, no new planets could be found; the number of planets remained fixed at six. Indeed, before 1772, astronomers did not make any serious attempt to discover new planets.

The first inkling that there might be more planets in the Universe came in 1772 when German astronomer Johann Daniel Titius made a remarkable discovery and which was popularized by another German astronomer Johann Bode. Johann Titius (January 2, 1729–December 11, 1796) was born in Konitz, Germany (now Chojnice, Poland). His father, a draper and also a city councilor, died when he was young, and he was brought up by his maternal uncle. His uncle encouraged his interest in natural science. After completing his school education, Titius joined the University of Leipzig and after four years of study received his master's degree in 1752. In 1756 he was appointed professor ordinarius[1] for lower mathematics at the University of Wittenberg and became a full professor in 1762. At the university, in addition to his courses in mathematics and physics, he also lectured on philosophy, theology, and law. While he was well versed with the astronomy of his time, he did not make any original contribution apart from the empirical law governing the distances between Sun and planets. The law which is known as Bode–Titius or Titius–Bode law is as follows:

> If one adds 4 to each term in the series, 0, 3, 6, 12, 24, 48, 96, 192, 384,…, and then divide it by 10, the result is approximately the average distances from the Sun to the then known planets in astronomical unit (AU) where the Earth's distance is taken as unity.

The law was popularized by the German astronomer, Johann Elert Bode (January 19, 1747–November 23, 1826). Son of a Hamburg merchant, Bode never attended formal schools and was taught by his father. He was deeply interested in mathematics and his prowess for calculation was phenomenal. In his youth, Bode suffered from eye disease, which, in particular, damaged his right eye. In later years also, he used to suffer from eye-related problems. When he

---

[1] In Germany a professor ordinarius has control over the teaching of his subject and also takes part in administration of the university.

was 18 years old, Bode came into contact with Johann Georg Büsch (1728–1800), a professor of mathematics. Greatly impressed by his mathematical prowess, Büsch allowed Bode to use his library and instruments for scientific self-education. In 1766, when he was 19 years old, Bode wrote a paper on solar eclipse, his first scientific contribution. In 1768, Bode published his book, *Instruction for the Knowledge of the Starry Heavens*. The book was very popular and in the second edition of the book Bode included the empirical law on planetary distances, originally found by J. D. Titius. With the book, the law also became popular. Bode became director of the Berlin Observatory in 1786 and made several important contributions in astronomy. He withdrew from official life in 1825.

Bode–Titius law was the first to gave the hint that solar system may contain other planets. In Table 4.1, planetary distances from Bode–Titius law are compared with the true values.

Known planets, Mercury, Venus, Earth, Mars, Jupiter and Saturn correspond to the 1st, 2nd, 3rd, 4th, 6th and 7th term in Bode–Titius

Table 4.1. Planetary positions in astronomical units from Bode–Titius law and actual positions are noted. The planets in parenthesis were not known when the law was given.

Bode–Titius series term	Distance from Bode–Titius law	Planet	Average distance
0	0.4	Mercury	0.39
3	0.7	Venus	0.72
6	1.0	Earth	1.0
12	1.6	Mars	1.52
24	2.8	—	—
48	5.2	Jupiter	5.20
96	10.0	Saturn	9.58
192	19.6	(Uranus)	19.19
384	38.8	(Neptune)	30.07
768	77.2	(Pluto)	39.49

lineup. One immediately notices the vacancy at the fifth position, no known planet corresponding to the 5th term at 2.8 AU. The vacancy gave the first inkling that solar system may contain more planets. Astronomers all over the world started looking for the seventh planet at 2.8 AU. While the search for the planet at 2.8 AU proved to be elusive, it resulted in discovering a new planet, Uranus. Uranus was discovered by Sir William Herschel (November 15, 1738–August 25, 1822) a German-born British (amateur) astronomer. Born in a poor family (his father was a musician in the Hanover military band), Herschel had to abandon schooling at the young age of fourteen. Trained in music, he followed his father to the military band. In 1759, following French occupation of Hanover, Herschel migrated to England, where he taught music before becoming an organist. Later, his sister, Caroline, joined him. Herschel's interest in astronomy grew late after his acquaintance with some English astronomers. Though Herschel did not have formal education, he had an insatiable thrust for knowledge and on his own learned mathematics and astronomy. He read Robert Smith's *Harmonics, or, The Philosophy of Musical Sounds* to learn about the theory of music. He read Robert Smith's *A Compleat System of Opticks*, and learned the techniques of telescope construction. From his urge to observe the sky, he made his own telescope, grinding and polishing mirrors, and started diligently looking up at the sky. Soon he was looking for the double stars (stars separated by a small distance). In the course of his observations, he noticed an uncommonly bright disc-like object. Initially he thought it to be a comet. He recorded his observations:

"On Tuesday, the 13th of March, 1781, between ten and eleven in the evening, while I was examining the small stars in the neighborhood of η Geminorum,[2] I perceived one that appeared visibly larger than the rest: being struck with its uncommon magnitude,

---

[2] Gemini is one of the 48 constellations (a set of stars) listed by the 2nd-century astronomer Ptolemy. H Geminorum is a twin star in the constellation. Auriga is another constellation.

I compared it to η Geminorum and the small star in the quartile between Auriga and Gemini, and finding it so much larger than either of them, suspected it to be a comet.

I was then engaged in a series of observations on the parallax of the fixed stars, which I hope soon to have the honor of laying before the R. S.; and those observations requiring very high powers, I had ready at hand several magnifiers of 227, 460, 932, 1536, 2010, etc., all of which I have successfully used on that occasion. The power I had on when I first saw the comet was 227. From experience I knew that the diameters of the fixed stars are not proportionally magnified with higher powers, as the planets are; I therefore I now put on the powers of 460 and 932, and found the diameter of the comet increased in proportion to the power, as it ought to be, on a supposition of its not being a fixed star, while the diameters of the stars to which I compared it, were not increased in the same ratio. Moreover, the comet being magnified much beyond what its light would admit of, appeared hazy and ill-defined with these great powers, while the stars preserved that lustre and distinctness which from many thousand observations I knew they would retain. The sequel has shown that my surmises were well founded, this proving to be the comet we have lately observed."

However, later measurements revealed that unlike a comet which moves fast, Herschel's object moves rather slowly. Also the object, unlike a comet, did not have a "coma" or tail, and soon, astronomers found its orbit to be nearly circular. Many more observations ultimately established the object as a planet well beyond Saturn. Herschel tried to name the new planet as 'Georgian star' (*Georgium sidus*) after his benefactor, King George III; however, after many deliberations, it was decided to continue the naming of planets after Greco-Roman Gods. Johan Bode proposed the name "Uranus" after the Greek God Uranus, father of Saturn and grandfather of Jupiter.

The discovery of Uranus made Herschel famous. He was knighted. He was also appointed a "Court Astronomer." The associated pension allowed him to leave his organist job. Later, he was elected as Fellow of Royal Society. Till his death, he devoted himself to astronomy, discovered thousands of nebulae (galaxies). His sister Caroline assisted him throughout his life. She herself discovered several comets and was the first woman to be paid for her contribution to science, to be awarded a Gold Medal of the Royal Astronomical Society and to be named an Honorary Member of the Royal Astronomical Society.

Herschel found Uranus at an average distance of 19.18 AU, further corroborating Bode–Titius law. However, the planet corresponding to the 5th position, at 2.8 AU was yet to be discovered. At the close of the 18th century, Hungarian astronomer, Baron Franz von Zach, organized a group of twenty-four astronomers to systematically search for the "missing planet." The group was called the "Celestial Police." In January 1, 1801, not the Celestial Police, but an Italian astronomer, Giuseppe Piazzi, working at the Palermo Observatory in Sicily, discovered an astronomical object at a distance of 2.77 AU. It was named Ceres after the Roman Goddess of agriculture and thought to be qualified to fill the vacancy in Bode–Titius lineup. However, soon it was discovered that Ceres had many sisters and could not be qualified to be a planet. Today, we know that it is an asteroid.[3]

In 1845, the English astronomer, John Adams, noted that the observed motion of Uranus showed discrepancy from model calculations. He reasoned that Uranus' path was perturbed by an eighth planet, yet to be discovered. He charted the course of the unobserved eighth planet. Independently of him, a French astronomer Urbain Le Verrier, also arrived at a similar conclusion and also charted the course of the yet to be discovered planet. In 1846, two German astronomers, Johann Galle and Heinrich d'Arrest found the eighth planet Neptune (named after the Roman God of the sea), at

---

[3]Asteroids are small (less than 100 kilometer in diameter) rocky, airless astronomical bodies revolving around the Sun. Asteroid belt between Mars and Jupiter, at 2.2 AU to 3.3 AU from Sun, contains thousands of asteroids with total mass less than that of Earth's Moon.

a distance of 30 astronomical units. Simultaneously, the winning streak of Bode–Titius law, which predicted the eighth planet at a distance of 38.8 AU came to an end. In the subsequent years, from precise measurements of planetary orbits and theoretical calculations, astronomers were certain of the presence of a ninth planet. However, it was much more elusive and only in 1930, American astronomer, Clyde Tombaugh, found the ninth planet. Following the tradition of Greco-Roman names for the planets, it was named Pluto after the Roman God of the underworld, who was able to render himself invisible.

From 1992 onwards, astronomers discovered a vast population of small bodies orbiting the Sun beyond Neptune. There are at least 70,000 trans-Neptunian objects with diameters larger than 100 km, confined within a thick band between 30 AU to 50 AU. They are called "Kuiper Belt objects." Incidentally, if Pluto was discovered now, it would not have been termed as a planet, rather a Kuiper Belt object. Indeed, at one time astronomers toyed with the idea of reclassifying Pluto as a minor planet or a Kuiper Belt object. But older generation learned from their school days about nine planets and they did not want to lose one of their planets. It is now reclassified as a dwarf planet. What is the difference between a planet and a dwarf planet? According to the world body, International Astronomical Union, a planet is a celestial body with three characteristics: (i) it orbits the Sun, (ii) it is sufficiently massive to be in hydrostatic equilibrium[4] and obtain nearly spherical shape, and (iii) it has cleared the neighborhood of its orbit. A dwarf planet satisfies the first two characteristics, but has not cleared its neighborhood. In 2004–2005, astronomers at Palomar Observatory, led by Mike Brown discovered several dwarf planets: Eris, Haumea and

---

[4]The principle of hydrostatic equilibrium can be defined for a fluid — a subset of matter, including liquid, gas, plasma and plastic solid. In hydrostatic equilibrium, the pressure at any point in a fluid at rest is just due to the weight of the overlying fluid.

Figure 4.1.   Our solar system. The image credit goes to NASA.

Makemake. Eris was named after the Greek Goddess of chaos, strife, and discord; Haumea with one-third mass of Pluto was named after the Hawaiian Goddess of childbirth; and Makemake was named after the God of fertility in Rapanui[5] mythology.

In Figure 4.1, our solar system as we know today is shown. Some physical characteristics of the heavenly bodies, e.g. mass, radius, distance from Sun, number of moons etc. are listed in Table 4.2. Mercury and Venus do not have any moon. Our planet Earth has a single moon or satellite: Moon. Mars has two moons: Phobos and Deimos. Galileo as early as 1610 discovered four moons of Jupiter: Io, Europa, Ganymede and Callisto. Later, astronomers discovered 63 more. Saturn, Uranus, Neptune and Pluto have 62, 27, 15 and 5 moons respectively. Also, between Mars and Jupiter lies the asteroid belt. Beyond the planet Neptune lies the Kuiper Belt, which is a region of space filled with trillions of icy

---

[5]The Rapanui people live on Easter Island in the southeastern Pacific Ocean. Easter Island is famous for Moai statues, the head and torso figures curved from stone.

Table 4.2. Some physical characteristics of our solar system.

Gravitational body	Radius (km)	Mass (kg)	Distance from Sun (km)	Number of Moons
Sun	695,500	1.99×10^{30}	—	—
Mercury	2440	3.30×10^{23}	57,910,000	0
Venus	6050	4.87×10^{24}	108,200,000	0
Earth	6378	5.97×10^{24}	149,600,000	1
Mars	3400	6.42×10^{23}	227,940,000	2
Jupiter	69,911	1.90×10^{27}	778,330,000	67
Saturn	58,232	5.68×10^{26}	1,424,600,000	62
Uranus	25,362	8.68×10^{25}	2,873,550,000	27
Neptune	24,622	1.024×10^{26}	4,501,000,000	14
Pluto	1184	1.31×10^{22}	5,945,900,000	5

cold bodies, revolving around the Sun. It is called "Kuiper Belt" in honor of the Dutch-American astronomer Gerard Kuiper. In 1951, Kuiper predicted the existence of a belt of icy objects at the outer solar system, remnants of the early solar system. The belt is much like the asteroid belt but instead of rocky objects, the objects here are icy cold. Scientists estimate that thousands of bodies of approximately 100 km in diameter, travel around the Sun within this belt, along with trillions of smaller objects, many of which are short-period comets with orbital periods of 200 years or less.

Of the eight planets in our solar system — Mercury, Venus, Earth, Mars, Jupiter, Saturn, Uranus, and Neptune — the first four are rocky, solid and are called "terrestrial" planets. They are composed essentially of silicate rock and metal. The rest, from Jupiter onward, are essentially big balls of gas. They are called "Jovian" (after the planet Jupiter) planets or gas giants. Jovian planets are not made entirely of gas; rather it is likely that they have a rocky core in the interior. The asteroid belt marks the transition point between the terrestrial and Jovian planets.

We have learned about our solar system. But how was it formed? Scientists have come up with a theory called solar nebular theory to explain the origin of the solar system. Understanding of the model will be facilitated with basic knowledge of atomic theory of matter, which is briefly discussed below.

4.2 What are we made of?

From the very beginning of civilization, men wondered about the constituents of the Universe. What are we made of? Greek philosopher Democritus, who lived during ca. 460–370 BCE, possibly pondered over the question: "Can one continue to divide a piece of matter indefinitely?" He concluded that no, it must end at some point when the matter could not be divided further into smaller pieces. Democritus called the smallest piece of matter *atomos*, meaning indivisible. From atomos we inherited the word atom. Democritus thought that our world was made of different kinds of atoms.[6] Greek philosopher Aristotle (384–322 BCE), on the other hand, believed that the entire world was made of four terrestrial roots or elements: earth (land), air, fire and water; and one extra-terrestrial element: aether. He did not believe in atomic theory and as his view dominated the western world, atomic theory fell out of favor.

The concept of the atom was revived in the early 19th century. Pioneering works of several chemists: Robert Boyle, Joseph Priestley, Henry Cavendish, Daniel Rutherford, Joseph Black, Antoine Lavoisier, Joseph Louis Proust and many others culminated in John Dalton's atomic theory of matter, who said *"all matter is made of atoms. Atoms are indivisible and indestructible."* In 19th century, in quick succession, scientists discovered several new phenomena which radically changed their concept of "what are we

---

[6] Hindu sage and philosopher, Kanad, who lived around 2nd century BCE (or alternately in 6th century BCE) also gave the concept of atom. He called them *"anu."*

made of?" German scientist Wilhelm Conrad Roentgen discovered the X-ray; English physicist Sir Joseph John Thomson discovered the electron, a negatively charged particle with a tiny fraction of the mass of hydrogen atom. Thomson also confirmed that electrons are part of the atom. French physicist Antoine Henri Becquerel, polish scientist Marie Skłodowska-Curie and her husband, French physicist Pierre Curie discovered radioactivity — spontaneous emission of ray from certain heavy elements like thorium, radium etc. New Zealand-born scientist Ernest Rutherford discovered the "nucleus," a small region where most of the mass of an atom is concentrated. Rutherford hypothesized that hydrogen nucleus is a fundamental constituent of all elements and gave it a special name "proton." Rutherford's student James Chadwick discovered the "neutron," a neutral particle that also exists in the nucleus. All these discoveries led scientists to revise their concept of the atom.

An atom has a positively charged core, called nucleus which is surrounded by negatively charged particles called electrons. Nucleus is made of positively charged protons and neutral neutrons, occupying only a tiny fraction of the atomic volume, yet containing most of the atomic mass. Since an atom is neutral it has an equal number of electrons and protons. A scheme was developed to symbolically represent each element: nucleus of an element can be designated as $^{A}_{Z}X$, where X is characteristic of the element, e.g. H for hydrogen, C for carbon, O for oxygen, Fe for iron etc. Z in the subscript is the number of protons in the nucleus and is called "atomic number." The superscript A is called the mass number and is the sum of proton and neutron numbers. Symbolic representations of few nuclei are given below:

$$\text{Hydrogen (1 proton + 0 neutron): } {}^{1}_{1}\text{H}$$

$$\text{Deuteron (1 proton + 1 neutron): } {}^{2}_{1}\text{H}$$

$$\text{Carbon (6 proton + 6 neutron): } {}^{12}_{6}\text{C}$$

$$\text{Nitrogen (7 proton + 7 neutron): } {}^{14}_{7}\text{N}$$

From antiquity, men knew about gravity, electricity and magnetism. In the 19th century, Scottish scientist James Clerk Maxwell showed that electricity and magnetism are not separate forces rather they are different manifestations of a single force called electromagnetic force. With the discovery of nucleus scientists learned about two new forces: strong nuclear force or strong force — the force that bind protons and neutrons in a nucleus; and weak nuclear force or weak force — the force that causes radioactive decay. How do these forces compare with each other? Among the four forces, gravitational force is the weakest. Indeed, it is very, very weak compared to the other forces. For example, the weak force exceeds the gravitational force by a factor of 10^{25} (one followed by 25 zeros), the electromagnetic force is stronger by a factor of 10^{36} and the strong force by a factor of 10^{38}.

For understanding the solar nebular theory, proton, neutron and electron as the fundamental constituents of matter suffice. Indeed, for a long time, till the advent of accelerators, it was thought so. In 1930s, physicists invented accelerators, where, using electric and magnetic field, charged particles can be propelled to great energy. Charged particles can be made to collide. In such energetic collisions, scientists discovered hundreds of new particles which are fundamental in the sense protons and neutrons are. Some of them were as heavy as protons and neutrons, some comparatively light, but heavier than electrons. The new particles were categorized into two types, medium mass particles as mesons (derived from Greek "*mesos*" meaning middle or intermediate) and heavy mass particles (including proton and neutron) as hadrons (derived from Greek "*hadros*" meaning bulky). In the process of understanding these large numbers of so-called fundamental particles — mesons and hadrons — scientists discovered "quarks," the truly fundamental or elementary particles. All the mesons and hadrons are made of quarks.

In the meantime, there was much progress in understanding electricity and magnetism. The concept of field was introduced. The concept of field is rather general. Anything varying over space and time can be considered as a field. For example, you are measuring the elevation at different places on the Earth. The elevation will vary with longitude and latitude. You can say elevation of places on the Earth is a field, more specifically, a two-dimensional field, since elevation varies with longitude and latitude. Michael Faraday introduced the concept of field to understand electricity and magnetism. In the language of field, an electric charge introduces its influence or field in the surrounding space. Depending on the nature of the field, another charge within the influence or field will either be attracted or repelled. As shown in Figure 4.2, Faraday visualized field as lines of force, emanating from a positive charge and terminating on a negative charge. Fields have an existence of their own. One can imagine lines of force even in a charge-free space when the field line forms a loop. Later, utilizing Faraday's concept of field and identifying the Faraday lines and loops with the electric and magnetic field, Scottish scientist James Clark Maxwell formulated his theory of electromagnetism, unifying electricity and magnetism.

In early 20th century, scientists also discovered that atoms and subatomic particles do not obey Newton's laws of motion, rather their behavior is dictated by a radically different law called quantum mechanics. Later, we will learn briefly about quantum mechanics.

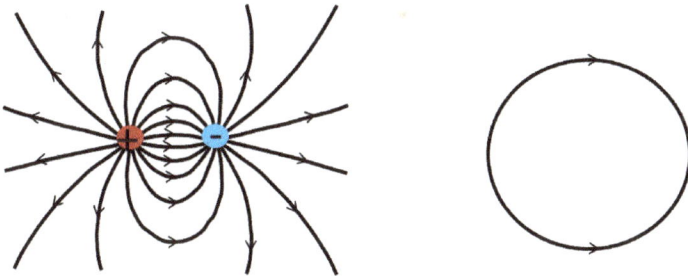

Figure 4.2. Left panel: The field lines between a positive and negative charge. The field lines emerge from the positive charge to terminate on the negative charge. Right panel: The field line in charge-free space. The field line terminates on itself forming a loop.

In the quantum version of Maxwell's theory of electromagnetism, which has been christened as quantum electrodynamics, fields and particles became synonymous. In this theory, two charged particles interact by exchanging another particle called photon.[7] Quantum electrodynamics has been highly successful in explaining all the aspects of electricity and magnetism.

In the 1970s, the idea of quantum electrodynamics was generalized to strong and weak nuclear forces to obtain the Standard Model of particle physics, which gives a consistent quantum mechanical description of the three basic forces: electromagnetic, strong and weak force. In the Standard Model, our world is made up of a handful of elementary or fundamental particles as listed in Table 4.3. Each elementary particle is characterized by mass, charge and a quantity called spin. The spin of an elementary particle is a purely quantum mechanical property, unrelated to the physical spinning of the body. An elementary particle can have either half-integer spin (1/2, 3/2, 5/2, …) or integral spin (0, 1, 2, 3, 4, …), but not in between, say 1/3 or 1/5. A half-integer spin particle is called a fermion (after the Italian scientist Enrico Fermi) and an integral spin particle is called a boson (after the Indian scientist Satyendra Nath Bose). Depending on the spin, elementary particles are classified as (i) matter particles, which are fermions; and (ii) mediator particles, which are bosons. The matter particles are of two types: quarks and leptons. Quarks are fractionally charged particles i.e. they contain a fraction of the electron charge and have a peculiar property that they do not exist freely — they can exist only in a meson or a hadron. In addition to mass, charge and spin, quarks are endowed with another property called color charge. Like

---

[7] Photon is an elementary particle, which made up all kinds of radiation including light. In physics, radiation is the emission or transmission of energy in the form of waves or particles through space or through a material medium. Visible light, radio waves, X-ray, gamma ray etc. are called electromagnetic radiations, which are propagations of oscillating electric and magnetic fields.

Table 4.3. List of elementary particles in the Standard Model. Mass, charge, spin and color charge of the particles are indicated along with their symbol. Mass is given in three units, eV/c^2, MeV/c^2 and GeV/c^2. $1\ eV/c^2 = 1.78 \times 10^{-36}$ kg. Charge is given in unit of electron charge.

		Name	Charge	Mass	Spin	Color charge
Matter particles	Quark	Up (u)	2/3	$2.3\ MeV/c^2$	1/2	yes
		Down (d)	−1/3	$4.8\ MeV/c^2$	1/2	yes
		Strange (s)	−1/3	$95\ MeV/c^2$	1/2	yes
		Charm (c)	2/3	$1.3\ GeV/c^2$	1/2	yes
		Bottom (b)	−1/3	$4.2\ GeV/c^2$	1/2	yes
		Top (t)	2/3	$173\ GeV/c^2$	1/2	yes
	Lepton	Electron (e)	−1	$0.511\ MeV/c^2$	1/2	no
		Muon (μ)	−1	$105.7\ MeV/c^2$	1/2	no
		Tau (τ)	−1	$1.8\ GeV/c^2$	1/2	no
		Electron neutrino (ν_e)	0	$<2.2\ eV/c^2$	1/2	no
		Muon neutrino (ν_μ)	0	$<2\ eV/c^2$	1/2	no
		Tau neutrino (ν_τ)	0	$<15.5\ eV/c^2$	1/2	no
Mediator particles		Photon (γ)	0	0	1	no
		Gluon (g)	0	0	1	yes
		Z boson (Z)	0	$91.2\ GeV/c^2$	1	no
		W boson (W)	0	$80.4\ GeV/c^2$	1	no
		Higgs boson (H)	0	$126\ GeV/c^2$	0	no

spin of an elementary particle, color charge of a quark is also of quantum mechanical origin and should not be confused with the ordinary color of an object. There are three types of color charges: red, green and blue (the names are so chosen as the combination of red, green and blue produces white color). Colored quarks combine to form a zero color charge hadron or meson. For each of these matter particles — quarks and leptons — there exists an antiparticle with the same mass and spin but opposite charge (including color charge), e.g. for an up quark with charge 2/3 e and color charge say red, there is an anti-up quark with charge –2/3 e and color charge anti-red; for an electron with charge –e, there is an antiparticle, called positron with charge +e.

Table 4.3 also lists the mediator particles. The mediator particles mediate force or interaction. Whenever two matter particles interact, a mediator particle is exchanged between them. The mediator particle decides the nature of the interaction. Thus, if W or Z bosons are exchanged, matter particles will interact by the weak force; if a gluon is exchanged, matter particles will interact by strong force; and if a photon is exchanged, matter particles will interact by electromagnetic force. How can exchange of a mediator particle give rise to interaction? Consider the following situation: two men are standing on two boats side by side. If the wind is neglected, on still water they will remain side by side. However, if the occupant of one boat throws a heavy ball to the other, his boat will recoil. If the occupant of the other boat catches the ball, his boat will also recoil. The net result of exchanging a ball between two boats is that the boats drift apart as if they are interacting repulsively.[8] Gravity is not included in the Standard Model, but all the matter particles interact gravitationally. Theoretically, the gravitational interaction is mediated by a hypothetical particle called graviton, a spin-2 bosonic particle.

4.3 Nebular theory

Nebular theory was first proposed by Emanuel Swedenborg (January 29, 1688–March 29, 1772) and later by Immanuel Kant (April 22, 1724–February 12, 1804) and Pierre-Simon Laplace. Swedenborg was a Swedish scientist, inventor and a Christian mystic. From his childhood, he was interested in mathematics and

---

[8] Unfortunately, similar simple analog for attractive interaction could not be found.

natural science. After graduating from the University of Uppsala in 1709, Swedenborg spent five years abroad, visiting England, France, and Germany. He had a purpose. Among the European countries, Sweden was then the least developed. Swedenborg wanted to learn science and bring back the knowledge to Sweden. While on visit, he used to rent rooms from craftsmen, clockmakers, cabinetmakers etc., focusing on learning applied sciences, intending to share his practical knowledge with the professors of his home-land. On returning home, he devoted his time to natural science and various engineering works. He also wrote profusely on science. He became renowned for his mechanical inventions. He can be credited with the first sketch of a "flying machine" (Leonardo da Vinci's manuscript, which contains the first sketch of flying machine, was not known till 19th century). Apparently, he made the following profound remark when one of his friends casted doubt about the possibility of flying,

"The art of flying is hardly yet born. It will be perfected and some day people will fly up to the Moon. Do we pretend to have dis-covered everything, or to have brought our knowledge to a point where nothing can be added to it? Oh, for mercy's sake, let us agree that there is still something left for the ages to come!"

In later years, he became a Christian mystic. His book *Heaven and Hell* caught the fancy of the Christian world and was translated into many languages. In 1734, Swedenborg published his book, *Principia* which contained a version of the nebular hypothesis. The planets are formed by matter cast off Sun and due to vortex[9] action move to their permanent positions. It was not much of a theory rather speculation. More improved version was given by the German philosopher, Immanuel Kant.

---

[9]A vortex is a region in a fluid with some rotational motion. Vortices can be seen if a fluid (e.g. water) is stirred.

Kant is considered as one of the central figures of modern philosophy. Kant's intellectual work easily justified his own claim to have affected a Copernican revolution in philosophy. His major work, *Critique of Pure Reason* greatly influenced 19th century philosophers like Sopenhauer and Hegel. The following quote from *Critique of Pure Reason* amply illustrates Kant's philosophy,

> "All our knowledge begins with the senses, proceeds then to the understanding, and ends with reason. There is nothing higher than reason."

Born in Königsberg, Prussia, in a modest artisan family (his father was a harness maker), he was baptized as "Emanuel." Later he learned Hebrew and changed his name to "Immanuel," literary meaning of which is "God is with us." He showed a great aptitude for study at an early age. In 1740, aged 16, he enrolled at the University of Königsberg, known as Albertina, and studied philosophy, science and mathematics. His father's death in 1746 interrupted his study and he spent six years as a private tutor to young children outside Königsberg. He finally returned to Königsberg in 1754 and in the following year, began teaching at the Albertina. For the next four decades Kant taught philosophy there, until his retirement in 1796 at the age of seventy-two. He never married and lived a simple, but highly disciplined life. There are stories that neighbors used to set their clocks by his walk; such disciplined was his life.

In 1755, Kant published his book, *In the General History of Nature and Theory of the Heavens*, in which he laid out the nebular hypothesis. Armed with Newtonian gravitational theory, Kant tried to explain not only the solar system but the entire Universe. In Kant's model, the solar system begins with a chaotic mixture of elements of differing specific densities. The mixture is acted upon by a combination of attractive and repulsive forces. The attractive force caused the elements to condense around central points, while a repulsive or centrifugal force acting on them created a sideways impetus that produced curvilinear or, more correctly, elliptical motions to the condensing bodies.

Laplace also forwarded a similar theory in his book, *Exposition of a World System,* published in 1796. His theory is based largely on the observation that all the known planets revolve around the Sun in the same direction. In his theory, Sun was originally a giant cloud of gas or nebula that rotated evenly. With time the gas contracted due to cooling and gravity. Contraction caused the gas to rotate faster. This faster rotation would throw off a rim of gas, which following cooling, would condense into a planet. This process would be repeated several times to produce all the planets.

Now pre-requisite of any scientific theory or model is that it is compatible with existing experimental data. What are the experimental data we have on our solar system? The following characteristics of solar system can be considered as experimental data:

(a) All planets orbit the Sun in the same direction as the Sun's rotation,
(b) All planetary orbits are confined to the same general plane, and
(c) Terrestrial planets form near the Sun, and Jovian planets are further out.

Any viable theory for formation of solar system must explain the above mentioned characteristics. Solar nebula model does explain above characteristics. Briefly, the model is explained in Figure 4.3.

The solar nebula is a giant cloud of gas and dust, which may be formed in a cosmic event like supernova explosion.[10] Now, we have

---

[10] A supernova is a giant explosion. When a star uses up its fusible material, it can no longer resist gravitational collapse. As the core of the star collapses, its density increases. The collapse continues till the density reaches 2–3 times the nuclear density. Thereafter, the core bounces back sending shock waves which completely disintegrate the core sending its enormous energy as radiation.

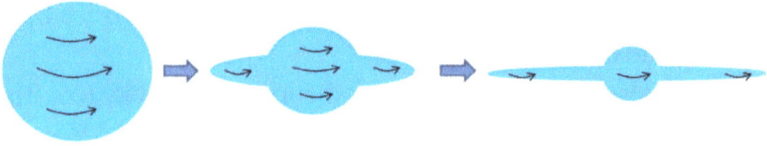

Figure 4.3.   Schematic diagram for the solar nebula theory.

a very good idea of the composition of the cloud. Approximately 70% of the cloud gas is neutral hydrogen, ionized hydrogen and hydrogen molecule, 20–25% is helium and the rest are various heavy elements in the gaseous phase. The dusts are essentially silicate grains (rock/sands) and graphite (carbon). There is no reason to believe that the gigantic cloud is created with no angular momentum (in physics, if a body moves in a straight line with velocity v, it is said to have linear momentum proportional to velocity. Similarly, if a body moves in a circle with some angular velocity, it is said to possess angular momentum, proportional to the angular velocity), so the giant cloud is probably spinning slowly. The cloud will be acted upon by two competing forces, an outward force due to gas pressure and an inward force due to gravitation. The outward gas pressure will try to expand the cloud; on the other hand, gravitational pull will try to contract it. When gravitational pull wins the competition, the cloud will start to contract. Now angular momentum has a property that it is conserved, i.e. once a body is given some angular momentum, it stays with it. Conservation of angular momentum now plays an important role. As the cloud contracts, conservation of angular momentum requires it to spin faster and faster. Because of the competing forces associated with gravity, gas pressure, and rotation (try to pull the cloud outwards), the contracting nebula begins to flatten into a spinning pancake shape with a bulge at the center (see Figure 4.3). The disc-shaped outer region is called protoplanetary disc. As the nebula continues to collapse, temperature of the central bulge increases. Gravitational potential energy is converted to the kinetic energy of individual gas particles falling inward. These particles crash into one another, converting their kinetic energy into heat. As the temperature increases, so does the gas pressure and the gas pressure starts to rival gravitational attraction; the collapse of the central bulge slows down and a proto-star is formed.

The young proto-star is a ball of hydrogen and helium not yet powered by fusion.[11] The proto-star continues to collapse and get heated up. Over the course of about fifty million years, the temperature of the material inside increases to million kelvin,[12] jump starting the fusion of hydrogen that drives the Sun today. The reactions proceed in three different stages. In the first stage, two hydrogen nuclei fuse to form a heavy hydrogen or deuteron (consisting of a proton and a neutron). The reaction is hard to take place. To fuse, two protons, which repel each other, have to come within 10^{-15} meters. This is only possible in very high temperature. Secondly, a proton has to convert to a neutron. This is a weak interaction process and the chance of occurrence is very small. The reaction rate is then very small (that is why the Sun is still burning even after 4.5 billion years). In the second stage, a hydrogen nucleus can get absorbed (or fuse) to a deuteron producing a helium-3 nucleus (with two protons and one neutron). In the third stage, two helium-3 nuclei can fuse to form helium-4 nucleus with the release of two protons. The two protons released in the third stage can again initiate the fusion process, setting up a chain reaction. The energy released in fusion process continues to heat the star. Like any heated object, the star also radiates energy. A stable star is formed when

---

[11] Fusion is a nuclear reaction when two or more of atomic nuclei come very close and collide at a very high speed and join to form a new type of atomic nucleus. Fusion of nuclei lower than iron generally releases energy, while fusion of nuclei heavier than iron absorbs energy.

[12] Kelvin is a temperature scale used in scientific discussions. The scale unit is same as that of the Celsius or centigrade scale with the identification 0 K = –273°C. 0 K, which can never be attained, is called absolute zero temperature, at which all molecular movement stops.

the energy radiated by the star equals the energy released from fusion reactions. In the three-stage fusion reactions described above, essentially 4 hydrogen nuclei fuse to form a helium nucleus. There is an alternate path, called CNO-cycle when carbon acts as a catalyst to fuse 4 hydrogen nuclei to helium. It is the dominant source of energy in heavy stars.

While the central bulge of the nebula converts into a star, the planets are formed from the gas and dust in the protoplanetary disc by a process called accretion. In the accretion process a massive object grows by gravitationally attracting more matter. Planets began as dust grains in orbits around the central proto-star. Through direct contact, these dust grains accumulate atoms to form larger bodies (planetesimals) of approximately 10 kilometers in size. Through further collisions the size gradually increases. Temperature across the protoplanetary disc is not uniform. Since different matters condensate at different temperatures, different types of planets are formed. Near the star, heavy elements, minerals condensate to form terrestrial planets. Further away, hydrogenous matter far outweighs heavy elements and mineral. They condensate to form the Jovian planets.

The solar nebula theory is consistent with the basic characteristics of the solar system. The orbits of the planets lie on a plane with Sun at the center. All the planets revolve around the Sun in the same direction, with the rotation axes nearly perpendicular to the orbital plane. It also explains why the Jovian planets are further away than the terrestrial planets.

4.4 Life and death of a star

Stars are born, live their life and die. Life span of a star depends on its mass. More massive is the star, less is its lifespan. When a star is born, it is powered by fusion energy. In a series of reactions, four hydrogen nuclei fuse to form helium nucleus. The energy released in the process counters the gravitational pull and the star is stable.

However, a stage will come when increased number of helium will interfere with hydrogen fusion, reducing its rate. The gravitational force will gain and the star will begin to contract. As the star contracts, heavy helium is pulled towards the center forming a core. Over the time as more and more helium nuclei are pulled towards the core, the core density and temperature increase. Hydrogen burning continues outside the core. At a certain stage, core density and temperature will increase to a point which enables helium nuclei to fuse in a series of reactions called triple-alpha (helium nucleus is called alpha particle) process and form carbon nucleus. Energy release from helium fusion, together with hydrogen fusion increase the thermal pressure, which overrides the gravitational pull and the star begins to expand. The star size increases to a great extent. The surface area increases even more and even though the star releases more energy from fusion processes, the surface actually gets colder and the star glows red. The star is now called a red giant star.

All the stars will reach the red giant stage. What happens later depends upon its mass. As the helium burning continues, more and more carbon nuclei are produced and get pulled towards the centre forming a carbon core. Outside the carbon core, helium and hydrogen burning continues. For an average-sized star, fusion process stops there as there is not enough gravitational force in the star to allow the temperature and density to reach the level where carbon nuclei can fuse and release energy. However, if mass permits, core density and temperature can increase to the extent that two carbon nuclei can fuse to form a neon nucleus and the associated fusion energy can power the star.

A carbon-burning star will be surrounded by helium-burning and hydrogen-burning shells. In the next stage, in the same process as the helium and carbon core was formed a neon core will form. In massive stars temperature and density of the neon core can be very high to initiate the photodisintegration reaction where a photon interacts with a neon nucleus to decompose it into oxygen and a helium nucleus. Over time, oxygen will drift towards the center and

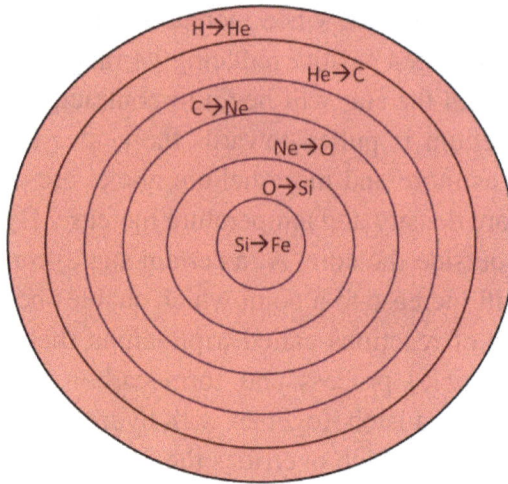

Figure 4.4.   Inside a fully developed star.

form an oxygen core. Subsequently, oxygen will fuse to form silicon core and subsequently silicon will fuse to form iron. Here the fusion process stops. Iron is the most stable nucleus and fusion of iron does not release any energy to power the star. As shown in Figure 4.4, a fully developed star will then have a core burning silicon to iron, surrounded by various shells burning hydrogen to helium, helium to carbon, carbon to neon, neon to oxygen and oxygen to silicon.

What happens to a star when all its fuel is burnt out? Fusion energy can no longer balance the gravitational pull and the star will start to contract. As the star contracts, density increases and at one stage, abundant electrons in the star will be so compressed that they will exert a pressure called degeneracy pressure. The origin of degeneracy pressure is quantum mechanical. Austrian-American physicist, Wolfgang Pauli made an important contribution in quantum mechanics by discovering a new law of nature. The law is now known as Pauli's exclusion principle. By exclusion principle, no two half-integer spin particles can occupy the same quantum state simultaneously. Electrons are half-integer spin particles. When electrons in the star are compressed to a small volume, they are forced to occupy higher energy quantum states and attain higher velocities, effectively leading to the degeneracy pressure. Before 1930, it was believed that for all stars, irrespective of their mass, the electron degeneracy pressure can balance the gravitational pull and the star will settle as a "white dwarf." At this stage the star is very dense and compact and shines with a white hot light. Once all of its energy is expended, the star becomes a black dwarf, where light no longer is emitted. However, time required even for a lowest mass white dwarf to cool to the extent that no light is emitted is much larger than the age of the Universe. Stars with mass ranging between 15–75 times the mass of Jupiter ultimately settle as brown dwarf. Hydrogen burning is not sustainable for stars in this mass range. Early in their life, they emitted visible light.

In 1930, a young Indian student was on a voyage to England. His name was Chandrasekhar Subramanyan (October 19, 1910–August 21, 1985) or Chandra to his admirers. Nephew of the Indian Nobel Laurate Chandrasekhara Venkata Raman, Chandra had his early education in Hindu High school, and Presidency College, Madras (presently Chennai). While in college he was inspired by a lecture of Arnold Sommerfeld, the German theoretical physicist, who made pioneering contributions in atomic and quantum physics. At that time, very few scientists had a good grasp of quantum mechanics. Still an undergraduate student, Chandra amply demonstrated his understanding of quantum mechanics by publishing two papers on Compton scattering,[13] one in *Indian Journal of Physics* and the other in *Proceedings of the Royal Society of England*. The Royal Society paper was communicated by the noted astronomer

---

[13] In 1923 Sir Arthur Holly Compton, an American physicist discovered that photon wavelength can be changed if scattered from electron. The effect is important as wave nature of photon alone cannot explain the effect. The effect is an experimental proof that photon has particle structure.

R.H. Fowler. The 1930 voyage was for higher studies under Fowler. While on the voyage, Chandra contemplated on the possibility of a very massive star becoming a white dwarf. Using Einstein's relativity, Chandra arrived at an unexpected and controversial result: gravitational pull of a star with mass more than 1.4 times the solar mass could not be balanced by the degeneracy pressure, it would not form a white dwarf, rather continue to collapse. Chandra's result was vehemently opposed by the then leading scientists since its acceptance would logically imply the existence of black holes (massive objects from which even light cannot escape), which were considered to be unphysical. The most vehement opposition came from the noted astronomer, Sir Authur Eddington. Eddington was instrumental in popularizing Einstein's relativity to the English-speaking world. He wrote and lectured profusely on relativity. Eddington was the best person to appreciate Chandra's theory, but he could not accept the reality of a black hole. At a conference in 1935, Eddington told his audience that Chandrasekhar's work *"was almost a reduction* ad absurdum *of the relativistic degeneracy formula. Various accidents may intervene to save a star, but I want more protection than that. I think there should be a law of nature to prevent a star from behaving in this absurd way!"* Chandrasekhar was deeply hurt by Eddington's comment and changed his research topic. However, much later, his theory was vindicated and he was awarded the 1983 Nobel Prize, *"for his theoretical studies of the physical processes of importance to the structure and evolution of the stars."*

Nowadays, scientists believe that life of a star ends as follows: towards the end of its life, when a star has used up most of its fuel, nuclear reactions become unstable. Sometimes the star burns furiously, other times slowing down. These variations cause the star to pulsate and throw off its outer layers. The thrown-off materials form the pre-planetary nebula, the source material for future star formation. What happens next depends on the size of the core. If the core mass is less than the Chandrasekhar limit, 1.4 times solar mass, the stars will ultimately settle as a brown dwarf. More massive stars will continue to collapse due to gravitational effect, until the density reaches the nuclear density, density of the nucleus of an atom, approximately 2.3×10^{17} kg per cubic meter. Further collapse will be resisted by the high density core and infalling matter will bounce off the core. This sudden core bounce will produce a supernova explosion. For a few weeks the star will burn brighter than a whole galaxy. Sometimes the core will survive the supernova explosion. If the star mass is less than ten solar mass, the core will survive as a neutron star. A neutron star is composed mainly of neutrons. All the protons in the star capture an electron to convert into neutron and the gravitational pull is counter balanced by neutron degeneracy pressure. For more massive stars, the neutron degeneracy pressure cannot balance the gravitational pull and the core will collapse into a black hole. The gravitational pull of the core will be so strong that even light cannot escape from its surface.

4.5 Black hole

The concept of the black hole was first introduced by British philosopher John Michell (December 25, 1724–April 29, 1793). Son of an English priest, Michell was educated at the Queen's College, Cambridge and later went on to teach there on various subjects: Hebrew, Greek, mathematics, and geology. The American Physical Society has described Michell as being "*so far ahead of his scientific contemporaries that his ideas languished in obscurity, until they were re-invented more than a century later.*" Michell's achievements were numerous. He has been called father of seismology and father of magnetometry. In 1750, he published *A Treatise of Artificial Magnets* where he described in detail, methods for producing artificial magnets, *much stronger than found in nature*. He also correctly proposed the "inverse square law for magnetic forces" (Newton proposed an inverse cube law). In 1755, he wrote a treatise *Conjectures concerning the Cause and Observations upon the*

Phaenomena of Earthquakes. Here he suggested that earthquakes spread as waves through the solid Earth and were related to the off-sets in geological strata now called faults. Following his works on earthquake he was made a member of the Royal Society.

However, the most important work of Michell was his proposal of the "dark star": stars which exist but cannot be seen, as no light can escape from them. At that time there were two theories for light: (i) Newton's corpuscular theory where light is made up of small tiny particles called "corpuscles" and (ii) wave theory of Dutch physicist Christiaan Huygens where light is a wave. Now we know that both theories are correct. In Newton's corpuscular theory, corpuscles travel in a straight line with a finite velocity. The concept is similar to the present concept of photon with a crucial difference that while corpuscles have mass, photons are massless. Michell was interested to find the mass of the stars. In 1783 he came up with an idea. Treating light as corpuscular particles, Michell reasoned that light emitted from a star would be pulled towards the star, which in effect would reduce its velocity and the change in velocity could be a measure of the star's mass. We now know that Michell was wrong, the velocity of light does not change, it is a constant. But by reason-ing so he came up with the idea of black holes or dark stars.

On Earth, if an object is thrown upward with a velocity called "escape velocity," it can overcome the Earth's attraction and escapes to infinity. Alternatively, an object will not be able to leave the Earth if its velocity is less than the escape velocity. Escape velocity depends on the mass of Earth, but not on the object's mass. Michell then correctly predicted that in a sufficiently massive star, light, treated as a corpuscular particle, would not be able to overcome the gravitational pull. No light can be emitted from the "dark star." He wrote to his friend Henry Cavendish,

> "If the semi-diameter of a sphere of the same density as the Sun were to exceed that of the Sun in the proportion of 500 to 1, a body falling from an infinite height towards it would have

acquired at its surface greater velocity than that of light, and consequently supposing light to be attracted by the same force in proportion to its *vis inertiae*, with other bodies, all light emitted from such a body would be made to return towards it by its own proper gravity. This assumes that light is influenced by gravity in the same way as massive objects."

Importantly, Michell also suggested a method to detect these invisible stars: astronomers should look for star systems that show the gravitational effects of two stars, and yet only one star is visible. This, he figured, would mean that the other star is a dark star. It was an extraordinary insight. Currently, there are over a dozen stellar black hole candidates in the Milky Way. Every single one of them is part of a so-called X-ray compact binary system.[14]

Twelve years later, Pierre-Simon Laplace proposed a similar idea, independently. Mathematics being his strong point, he could also provide a mathematical proof. He wrote:

"The gravitation attraction of a star with a diameter 250 times that of the Sun and comparable in density to the Earth would be so great no light could escape from its surface. The largest bodies in the Universe may thus be invisible by reason of their magnitude."

It may be worth mentioning that Michell's or Laplace's concept of a black hole is far from the present day concept. Both Michell and Laplace increased the gravitational pull by increasing the size of the stars, while keeping the density same as in normal stars. In modern concept however, gravitational pull is increased by making the stars infinitely dense.

---

[14] X-ray compact binaries are a class of binary stars where matter is transferred from one (a normal star) to the other (a compact object e.g. a white dwarf, neutron star or a black hole). The gravitational energy of the falling matter is released as X-ray.

The idea of the dark star was revived a century later when Albert Einstein (March 14, 1879–April 18, 1955) proposed his theory of general relativity. Born in Germany, off middle-class Jewish parents, Einstein is inarguably the most influential scientist of the 20th century. Two childhood events had marked effects on his development. At the age of five, he got hold of a compass. He marveled at the invisible force that turned the needle. The second, at the age of 12, he received Euclid's *Element* as a gift from his teacher. Again he marveled at the logic of arguments. Einstein received his primary education at a Catholic school and later at Luitpold-Gymnasium[15] in Munich. However, he did not like the rigid atmosphere at the gymnasium and left the school in 1894, without a degree. Later, he continued his study in a Swiss gymnasium at Aarau and obtained his school leaving certificate in 1896. He then enrolled in the four-year mathematics and physics teaching diploma program at the Zürich Polytechnic and completed the program in 1900. In the meantime, he renounced his German nationality and from 1896 to 1901 (till he became Swiss national) was a stateless person. After his graduation he had trouble finding a suitable job. In 1901 through the influence of one of his friend's father he secured a job in the Swiss patent office. In 1905, Einstein submitted a 24-page dissertation, "A New Determination of Molecular Dimensions" for the doctoral degree from the University of Zurich. Einstein recounted that his supervisor, Alfred Kleiner returned the thesis saying it was too short. Einstein added a single sentence and it was accepted without further comments. In 1905, 26 years old Einstein, a clerk at the Swiss patent office, published four papers in the German journal *Annalen der Physik*, each of which contributed enormously to the foundation of modern physics and changed the world view on space, time, mass, and energy. Such was the impact of Einstein's 1905 papers that physicists call the year *Annus Mirabilis*

---

[15] Gymnasiums are kind of secondary schools, with strong emphasis on academic learning, preparing students for university education.

(the wonderful or miraculous year). The four papers established the patent office clerk as a leading scientist. In 1908 Einstein left his job at the patent office and joined the University of Bern as a lecturer. In 1914, he was appointed Director of the Kaiser Wilhelm Institute for Physics (1914–1932) and a professor at the Humboldt University of Berlin, with a special clause in his contract that freed him from teaching obligations. In 1921, he received the Nobel Prize in Physics for the explanation of the photoelectric effect.[16] In 1933, when Hitler assumed power and started persecuting Jews, Einstein left Germany and joined Princeton University in USA, where he continued until his death.

In one of his 1905 papers, Einstein proposed the theory of special relativity. The theory is based on two key postulates: (i) the laws of physics are same in all inertial (constant velocity) frames and (ii) the speed of light (generally denoted by the symbol c) in vacuum is a constant. The theory is special because it deals only with constant-velocity inertial frames. Later he gave the theory for accelerating frames, which was christened as the general theory of relativity. The theory of special relativity generalizes the Newtonian mechanics of a small-velocity particle or a system of particles or fluid to velocity comparable to the velocity of light. Indeed, Newtonian mechanics is a special case of relativistic mechanics with small velocity. The theory sprang a few surprises. For example, in Newtonian mechanics, two velocities v_1 and v_2 are added up to $v = v_1 + v_2$. The same in relativistic mechanics is,

$$v = (v_1 + v_2)/(1 + v_1 \times v_2/c^2).$$

When compared to velocity of light (186,000 mile/second), velocity we encounter in everyday life is negligibly small and the second term in the denominator can be safely ignored. Relativistic addition

---

[16] In photoelectric effect, certain metals when shined by light emit electrons. Einstein, using quantum mechanics gave proper explanation of the phenomenon. Indeed, it was the first application of quantum mechanics in a physical process.

formula and Newton's formula then become identical. Relativistic effect starts to manifest only when the velocities are comparable to velocity of light. Consider a case when both v_1 and v_2 are equal to velocity of light c. In Newtonian mechanics, two light velocities will add up to $2c$ but in relativistic mechanics, they will add up to c only.

The special theory of relativity radically changed our view of space and time. In classical or Newtonian physics, space and time have separate identities. Newton wrote:

"Absolute space, in its own nature, without relation to anything external, remains always similar and immovable;"

and

"Absolute, true and mathematical time, of itself, and from its own nature, flows equably without relation to anything external."

In special relativity, Einstein proposed that space and time are not separate entities, rather they are but one entity called space-time continuum. It is a continuum because according to our experience, there is no void in space or in time. In a public lecture on relativity, Hermann Minkowski, Einstein's former teacher, said:

"The views of space and time which we wish to lay before you have sprung from the soil of experimental physics, and therein lies their strength. They are radical. Henceforth, space by itself, and time by itself, are doomed to fade away into mere shadows, and only a kind of union of the two will preserve an independent reality."

Intermingling of space and time also sprang a few surprises. For example, a clock will tick slowly for an observer moving at the speed of light or close to the speed of light. This gave rise to the famous "twin paradox." Say, one of the identical twins embarks on a space journey on a rocket travelling near the speed of light. For him the clock will tick slowly. When he returns back to Earth, he will be much younger than his brother who remained on the Earth and for whom, the clock ticked at the normal pace. Another surprising feature is the length contraction. To a stationary observer, an object moving at relativistic speed will appear to shorten in length (along the direction of motion). Special relativity also changed our view about mass and energy. In Newtonian mechanics, they are two distinct entities. However, Einstein showed that they are indistinguishable, mass and energy is equivalent. Possibly, the most well-known equation in the world is Einstein's famous equation, $E = mc^2$ — mass times the square of speed of light is energy. In a sense, the equivalence of mass and energy led to the construction of the most lethal, destructive weapon known to mankind, the "atom bomb."

In 1915, Einstein published his theory of gravity, which he called general relativity. Einstein's view of gravity is radically different from that of Newton. According to Newton, gravity is the force of attraction between two massive objects. It is the same force that holds planets orbiting around the Sun as well as the force that pulls a falling apple to the Earth. He laid down the rule: the force (F) between two objects of mass M_1 and M_2 at a separation of R is governed by the inverse square law,

$$F = GM_1M_2/R^2,$$

where G is a constant called the gravitational constant.

In Newtonian mechanics, mass tells gravity how much force to exert and force tells the mass how to move. Unlike in Newtonian mechanics, in Einstein's relativity space and time are mixed up, they do not have separate existence. In the presence of mass (or equivalently energy), space-time gets curved (see Figure 4.5). The curved space-time tells the mass how to move. Gravity is nothing but curvature of the space-time. For example, if asked, why the Moon does not fly off into space, rather than orbiting around the Earth, Newton would have answered that the force of gravity acting

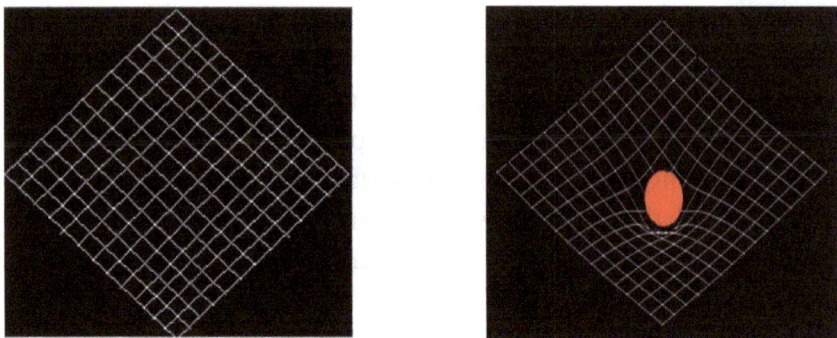

Figure 4.5. Gravity in Einstein's general relativity. In the absence of a mass space-time is flat (left panel). Space-time is curved in the presence of mass (right panel).

between the Earth and Moon hold it in the orbit. On the other hand, Einstein would have replied that the Earth's mass bends the space and time around itself and Moon follows the curves created by the massive Earth.

Einstein's general relativity had a specific prediction: light can bend in vicinity of a massive object like the Sun. Around a massive object, space-time is curved and light will follow the curved space-time, effectively departing from its straight line path. Einstein calculated the amount of bending due to the Sun. In 1919, during a total solar eclipse, Sir Arthur Eddington and his collaborators measured the bending of light, exactly matching Einstein's prediction. Reportedly, Einstein was asked what would have been his reaction, if experiments failed to detect the bending, Einstein famously quipped: *"Then I would feel sorry for the dear Lord. The theory is correct anyway."* Einstein's prediction of bending of light also posited that distributed matter can act as a gravitational lens. Consider a massive object lying between a star and an observer. Light from the star will bend due to gravitational effect. As shown Figure 4.6, bending of the light produces two images of the star. From the image separation, it is possible to infer about the mass of the object, luminous or dark. In reality, unlike the plane of the paper, space is

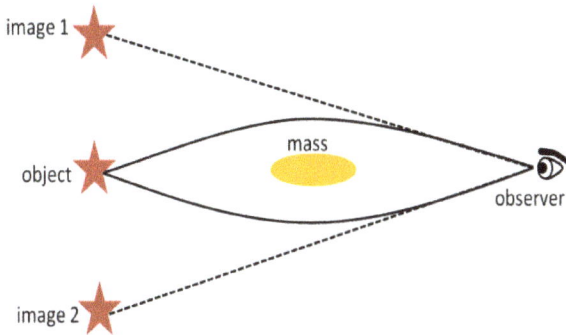

Figure 4.6. Schematic diagram showing gravitational lensing effect.

of three dimensions and instead of two light sources, the observer will see a ring of light sources. Gravitational lensing is now used in measurements of distant galaxies, shape of stars etc.

Einstein also predicted gravitational waves. Gravitational waves are ripples in the space-time fabric. Einstein's mathematics showed that massive accelerating body (such as neutron stars or black holes orbiting each other) will produce disturbances in the space-time fabric and the disturbance will propagate as a wave with speed of light, much like the wave produced in a still pond if water is disturbed by throwing a pebble. In 2016, approximately 100 years after the prediction, gravitational waves sent off from a merger of two massive black holes was discovered by the LIGO (which is acronym for Laser Interferometer Gravitational-Wave Observatory) collaboration.

Unlike the special relativity theory, Einstein's general relativity involves rather complex ideas and abstract mathematics. When it was published, it was rumored that only three people in the world understood it. Once, Sir Arther Eddington gave a seminar on relativity at the Royal Society, England. Eddington was instrumental in popularizing Einstein's work to the English-speaking world. After the seminar, Ludwik Silberstein, a well-known relativist and author of a decent treatise on special relativity asked Eddington, *"Isn't it true, my dear Eddington, that only three persons in the world understand relativity?"* Silberstein possibly thought that Eddington would say,

"You are one of them." However, Eddington did not reply. Silberstein insisted: "*Professor, you must be one of the three persons in the world who understand general relativity.*" To which Eddington, unruffled, replied, "*On the contrary, I am trying to think who the third person is!*"

In 1915, Karl Schwarzschild (October 9, 1873–May 11, 1916) wrote two papers on Einstein's relativity, which formed the basis for later studies of dark stars or black holes. Born in a Jewish business family in Frankfurt, Schwarzschild had his early education in a Jewish primary school and later at the University of Strassburg and Munich. He was a kind of child prodigy and before he was 16 years old, wrote two complex papers on astronomy. After receiving his doctorate degree, he joined an observatory in Vienna. Later, he became a professor at the University of Gottingen. He volunteered for military service during the first World War. In 1915, while on military service, he exactly solved Einstein's equations. He sent the solution to Einstein. Einstein replied:

> "I had not expected that one could formulate the exact solution of the problem in such a simple way."

Schwarzschild's solution has a spectacular feature. For a massive object, one can have a space-time region, where a space traveler could go all the way in, but never to return. Indeed, light also would not emerge out. Princeton physicist, John Archibald Wheeler dubbed these strange solutions as "black holes." Schwarzschild's solution also gave the condition under which a gravitational body will be a black hole. If G is the gravitational constant and c is the velocity of light, a body of mass M contained within a radius R_{sh} (called Schwarzschild's radius),

$$R_{\text{sh}} = \frac{2GM}{c^2} \approx 2.95 \frac{M}{M_{\text{sun}}} (\text{km}),$$

will behave as a black hole. If our Sun, whose radius is approximately 100 times the radius of Earth, can be compressed to a body of radius of ~2.95 km, it will become a black hole. Incidentally, Schwarzschild himself as well as Einstein and many leading scientists did not believe the physical reality of the solutions. They thought that the "black hole" solutions were due to incomplete knowledge and with better understanding they would eventually be removed. But over time, it was realized that Schwarzschild's solutions are real. The circumstances under which black holes are formed are now understood and it is believed that most of the galaxies contain a black hole in the deep interior.

Black holes are dark objects. How do we know that they are physical reality? Their presence can only be indicated by gravitational effect. A black hole will gobble up a nearby star. As the matter falls into the black hole, it can radiate X-rays, which can be detected. In 1972, Charles Bolton, an American astronomer, presented the first irrefutable evidence of existence of a black hole. As a young post-doctoral student at the David Dunlap Observatory at Toronto, he was studying binary stars. In binary stars, two stars rotate around their common center of mass.[17] Astronomers are interested in binary star systems as their mass can be calculated easily using Newton's gravitational law. Bolton detected a star wobbling, as if it was orbiting around a massive but invisible star emitting powerful X-rays. Calculations indicated that the invisible star is too heavy to be a neutron star. The invisible star is now recognized as the black hole Cygnus X-1.

---

[17] Center of mass is a point where all the mass can be assumed to be concentrated for theoretical calculation. For example, for a rod of constant density, the midway point will be the center of mass; for a sphere of constant density, the center of the sphere will be the center of mass.

Classically, a black hole is a simple system. There is a theorem, curiously named no-hair theorem, which states that a stationary[18] black hole is completely characterized by only three quantities: mass, electric charge and angular momentum. It has no temperature. In physics, there is a quantity called entropy, which measures the disorder or randomness in a complex system. The more disorder there is in a system the more is its entropy. In a way, entropy measures the number of ways the constituents of the system can be rearranged without changing the overall appearance. Entropy is zero for a zero-temperature system and classically, black holes do not have any entropy. Now, there is a well-tested law that states that entropy of a closed or isolated system can never decrease, it can only increase. A body like a star, with a large number of constituents, can have large entropy. If it collapses to form a black hole, eventually, by no-hair theorem it will have zero entropy. The entropy of the star will be lost; the well-tested entropy law will be violated. In 1973, Jacob Bekenstein, a student of John Wheeler, argued that black holes must have entropy. He proposed a startling formula: entropy of a black hole is proportional to the area of the horizon. It is now known as Bekenstein–Hawking formula. Black hole entropy raised a problem. How can a zero-temperature black hole have entropy? The resolution came with Hawking's discovery of black hole radiation.

4.6 Black holes are not so black

Nothing comes out of a black hole. However, in 1975, Stephen Hawking published an astounding result: if one takes into account quantum mechanics, black holes are no longer black, rather they emit radiation — blackbody radiation. In physics, radiation is the emission or transmission of energy in the form of waves or particles through space or through a material medium. Visible light, radio

---

[18]A stationary system remains unchanged with time. A stationary black hole remains the same as time elapses.

waves, X-ray, gamma ray etc. are called electromagnetic radiation as oscillating electric and magnetic fields propagate. Blackbody radiation is electromagnetic in nature and consists of all ranges of frequency or wavelength. A body, held at a constant temperature emits blackbody radiation.

Born on January 8, 1942, English theoretical physicist, Stephen Hawking is possibly best known for his hugely popular book, *A Brief History of Time*. While in school, Hawking did not show any sign of scholarship. However, he did develop a love for mathematics and decided to study mathematics. His father, himself a medicine man, considering few job opportunities for the mathematics graduates, wanted Hawking to study medicine. A compromise was reached and Hawking joined Oxford University College for a degree in physics. Hawking found academic works rather easy and by his own admission, spending one hour a day was sufficient for him to obtain the degree with honors. Later, in 1962, he joined Cambridge University for a Ph.D. in cosmology. While in Oxford, Hawking suffered from some physical problems, like slurring of speech, occasional tripping etc. The problems aggravated with time and in 1963 subsequent to several medical tests, doctors diagnosed his disease as the early stages of amyotrophic lateral sclerosis. It is a rare disease where slowly the nerves that control the muscles die. Doctors gave him two and a half years to live. The devastating news changed Hawking completely. Earlier, he did not find anything worthwhile to do. Now with his time limited, Hawking had a purpose, to finish his Ph.D. He devoted himself completely to his work and study. Before he was diagnosed with the disease, Hawking was in love with an undergraduate student, Jane Wilde. He married her in 1965. The couple was blessed with a son in 1967 and a daughter in 1970. In the course of time, Hawking's condition worsened, he was forced to use a crutch, later a wheelchair and his speech became completely unintelligible. He lost his capacity of writing also. However, physical disabilities could not dominate him and he obtained his Ph.D. in 1966. Overriding his progressive disabilities, he continued to work on various significant problems of cosmology and in 1979, he was appointed Lucasian Professor of Mathematics at Cambridge, which he held until 2009. According to him, he is foremost a scientist and then a writer, never a disable man.

To understand the Hawking radiation, we have to know about quantum mechanics. The concepts of quantum mechanics are rather difficult to grasp as they defy our everyday experiences. It is the theory of microscopic objects i.e. of atoms and subatomic particles. The theory states that the physical laws for microscopic objects like atoms etc. are radically different from the physical laws for macro-scopic objects. For example, a quantum particle can exist as a par-ticle as well as a wave. This is called wave–particle duality. Thus an electron which is a particle can exhibit wave–like properties and can cause interference.[19] In classical mechanics[20] a body or an object has a definite state of existence, we can say at this time the body is here, but not so for a quantum body. At any time, a quantum body can be everywhere and one can assign only a probability for its existence at any particular position. This is done by assigning a mathematical function[21] $\psi(x)$. $\psi(x)$ is called a wave function and at each instant can have non-zero value at each spatial position x. All the necessary information about the quantum body is encoded in its wave function. In other words, wave function fully specifies the state of a quantum body. A wave function is characterized by certain numbers called quantum numbers, which can have only discrete values. A peculiar feature of a quantum body is that at any instant it can exist in several quantum states and only a measurement can tell

---

[19] Interference is the phenomenon in which two waves combine to reinforce or cancel each other.

[20] Classical mechanics is the study of the motion of bodies in accordance with Newton's laws of motion.

[21] In mathematics, a function is a relation between a set of inputs and outputs. For example, say the set of inputs are real number, 1, 2, 3, ... , and the set of output is the square of the input, 1, 4, 9, ... The relation can be formally expressed as a function, $f(x) = x^2$, x = 1, 2, 3....

us in which state it is. The Schrödinger's cat (thought) experiment amply illustrates the situation. Say, in a closed box, a cat is put together with a vial of hydrogen cyanide and a small amount of radioactive material. By a special arrangement, whenever the radioactive material decays the vial will release the poison, killing the cat. The cat's life thus depends on whether or not the radioactive material decays; it will be dead if it decays and alive if not. Alive and dead are the two possible states for the cat. Quantum mechanically, until a measurement is done, i.e. the box is opened, the cat can be in both states with equal probability — simultaneously alive and dead, a situation defying our everyday experiences. Only a measurement will fix in which state it is.

Another intriguing law associated with quantum mechanics is that unlike in Newtonian mechanics where a body can have definite position and velocity (or momentum — mass times the velocity), a quantum body cannot possess definite position and velocity simultaneously. If the position of the body is measured with 100% certainty, its velocity or momentum will be infinitely uncertain. Alternatively, if the velocity or momentum of the body is measured with 100% certainty, its position again will be infinitely uncertain. German physicist Werner Heisenberg was the first to formulate this principle and it is now called Heisenberg's uncertainty principle. Mathematically, the uncertainty principle is written as,

$$\Delta x \, \Delta p \sim \hbar/2,$$

where Δx and Δp are uncertainties in position and momentum measurements and $\hbar$ is the reduced Planck's constant, $\hbar = h/2\pi$. In another variant of Heisenberg's uncertainty principle, energy and time are similarly related. Precision measurement of energy of a quantum system will require infinite amount of time and *vice versa*. Mathematically, if ΔE and ΔT are the uncertainties in energy and time, they are related as,

$$\Delta E \, \Delta T \sim \hbar/2.$$

One important consequence of Heisenberg's energy–time uncertainty relation is that in a short time span $\Delta T \sim \hbar/2\, \Delta E$, we can create and annihilate a particle–antiparticle pair with energy ΔE. The process is transient and the particles are virtual in the sense that they are not measurable and need not satisfy the energy or momentum conservation laws as required for the real particles.

The quantum mechanical possibility of producing virtual particles drastically alters our concept of vacuum. Classically vacuum is considered as an empty space. However, quantum mechanically vacuum is not an empty space; rather it is seething, continually creating and annihilating particle–antiparticle pairs, constrained only by the Heisenberg energy–time uncertainty principle.

With this limited introduction to quantum mechanics, let us try to understand Hawking radiation from a black hole. In Einstein's general relativity, gravity is nothing but the curvature of the space-time. If at some point space-time curvature becomes infinite, the gravitational attraction becomes infinite too. The point is then a gravitational singularity. In modern theories, black holes are gravitational singularities. The Schwarzschild radius is the horizon of the black hole, which acts as a one-way membrane — things can go in but do not come out. Now consider a black hole in an empty space. Say a virtual particle–antiparticle pair is created far outside the horizon. Within the timescale permitted by Heisenberg's uncertainty principle, the pair will be annihilated as well. The situation is pictorially shown in Figure 4.7, event-I. Now imagine a situation where

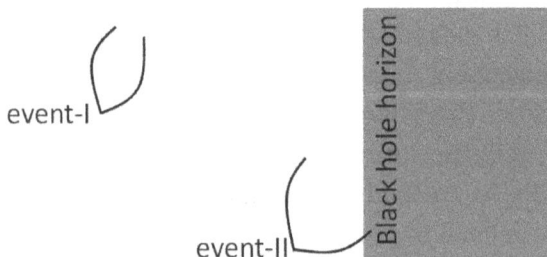

Figure 4.7.   The Hawking mechanism of black hole radiation is explained.

the pair is created near the horizon (event-II in Figure 4.7). One of the pair moves towards the event horizon, the other away from it. The two particles will face drastically different gravitational pull. The particle moving towards the horizon, once it crosses the horizon can never come out. Its partner particle moving away from the horizon cannot be annihilated. It will be realized as a real particle. The energy required for realizing the particle then comes from the black hole. It will appear as if the black hole is radiating a particle.

Black hole radiation solved the black hole entropy problem. Hawking showed that the radiation from a black hole is just like the blackbody radiation implying that black hole has a temperature. The temperature of a black hole can be calculated, it is inversely proportional to the black hole mass; the more massive is the black hole, the less is its temperature. The behavior is in contrast to our everyday experience. Normally, when fuel is added, temperature increases. For a black hole, adding fuel (mass) decreases its temperature. Hawking radiation also implies that black holes die. Over time, a black hole will radiate all its energy or mass and cease to exist. The lifetime is rather long, for example, a black hole with the mass of our Sun will evaporate in a time scale of 10^{67} years (note that our Universe is only 13 billion years old). Though the entropy problem was solved, black hole radiation gave rise to a new problem — the famous information paradox. There is a quantum mechanical law that states that information cannot be lost. Say a black hole is formed from a heavy star running out of fuel. Quantum mechanically, the black hole stores the information of the star that gave its birth. If the black hole does not radiate, one can think that the information is locked in the black hole. However, if the black hole radiates and ultimately evaporates, all that will be left is featureless radiation, carrying no information — apparently the information is lost. Though there is much progress in understanding black hole physics, the issue of information loss is not resolved yet.

Chapter 5

Contemporary Cosmology

> Learn from yesterday, live for today, hope for tomorrow. The important thing is not to stop questioning.
>
> Albert Einstein

5.1 We are expanding

To the ancient Greeks, our solar system was the entire Universe. The Greeks saw the patches of dim glowing light arcing across the night sky, they call it *Via Lactea* or Milky Way. While some Greek philosophers believed that Milky Way is a vast collection of stars, Aristotle, whose view dominated until the 15th century, believed it to be an earthly phenomenon. Galileo Galilei, in 1610, first resolved the band of light into individual stars with his telescope. The Universe then expanded to Milky Way and for a long time, it was thought that the entire Universe consists of Milky Way and our solar system is one among billions of such systems (though Earth may be a very special planet harboring intelligent life). It took more than three hundred years to know that Milky Way is not an isolated galaxy and the Universe contains billions of galaxies like Milky Way.

The term "galaxy" was coined by the English astronomer, Thomas Wright (September 22, 1711–February 25, 1786). Son of a yeoman and carpenter, Thomas suffered from a severe speech impediment and could not complete his schooling. At the age of 14, he became an apprentice to a watchmaker. However, at home, he continued his study of arithmetic and astronomy. Such alarming was his dedication to the study, that once, to stop him from reading, his father burned all his books. However, burning of the books could not dampen his urge for knowledge and over the years, he became a master of astronomy, mathematics, navigation, instrument making, architecture, and even gardening. He left the apprenticeship for a brief stint as a sailor. Later, he started teaching navigation to sailors. By that time, the speech impediment must have been reduced appreciably because he became popular as a teacher. Subsequently, he set up a school to teach mathematics and navigation. He wrote regularly on the solar eclipse, lunar eclipse, navigation and produced almanacs.[1] Thomas was disturbed by the Milky Way as it required that the stars are not evenly distributed in the Universe. A religious man, Thomas believed that God created a perfect Universe and stars should be evenly spaced in the Universe. He tried to model a Universe, where stars are evenly distributed yet may appear to be uneven to an observer on the Earth. He knew of the Saturn rings. He then proposed a disc-shaped galaxy. In a disc-shaped galaxy, even if the stars are evenly spaced, to an observer on the Earth, it would appear an uneven distribution. In 1750, Thomas' theory was published in his most influential book, *An Original Theory on New Hypothesis of the Universe.*

William Herschel (famous for discovering the planet Uranus), read a summary of the book. He decided to count the number of stars in a statistical way. He randomly selected 683 small regions

---

[1]An almanac is an annual calendar containing important dates and statistical information such as astronomical data and tide tables.

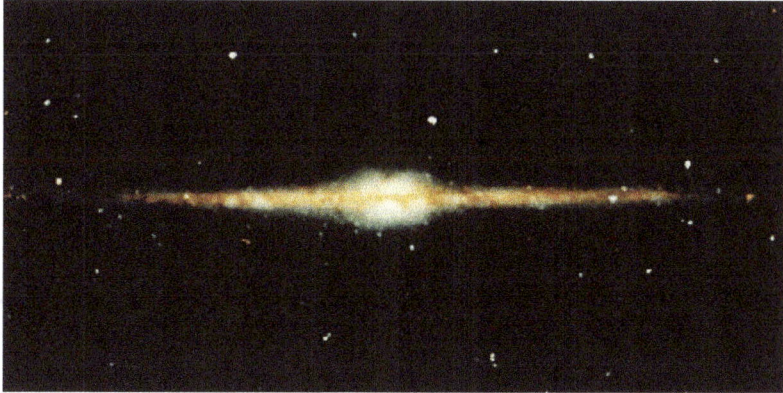

Figure 5.1. Image of the Milky Way. Image credit goes to NASA.

of the sky and with his 48-inch telescope, started the painstaking process of counting. He found that the number of stars per unit area increased as he approached the Milky Way and was minimum at the right angle to Milky Way. His counting established Thomas' hypothesis that we are located in a disk of stars, with the plane of the disk aligned with the hazy Milky Way. In recent years, our galaxy has been imaged. In Figure 5.1, infrared image of the Milky Way, taken by NASA's Cosmic Background Explorer (COBE) satellite is shown, visually proving Thomas' conjecture of our disc-shaped galaxy.

Until the early 1920s, it was believed that all the stars in the Universe are contained in the Milky Way, or Milky Way is our Universe. The view was changed by Edwin Powell Hubble (November 20, 1889–September 28, 1953), an American astronomer. Hubble revolutionized the extragalactic astronomy by conclusively proving that Milky Way is just another galaxy among billions of galaxies. In his younger days, Hubble was more noted for his athletic ability than his scholarship. He studied astronomy and mathematics at the Chicago University and obtained a bachelor's degree. Bowing to his parents' wish he studied law at Oxford, obtained the degree and returned back to Chicago. However, he was not interested in practicing law, and after his father's death joined the Chicago University for a Ph.D. degree in astronomy. In 1917, the year the United States declared war on Germany, he obtained his Ph.D. degree. George Hale, the founder of Mount Wilson Observatory,[2] offered Hubble a staff position at the observatory. However, Hubble volunteered for the military service in World War I, and sent a telegram to Hale, *"Regret cannot accept your invitation. Am off to the war."* However, the offer remained and after the war, Hubble joined Mount Wilson Observatory. At that time, Mount Wilson Observatory had installed the then world's largest 100-inch "Hooker" telescope, funded by the Los Angeles business man John Daggett Hooker.

In 1917, the most pressing question was the nature of the cloudy patches called nebulae.[3] Most of the astronomers thought them to be contained within our Milky Way galaxy. However, Hubble was not sure; he turned the Hooker telescope, the best then available, to those nebulae and started measuring their distances. Now, while the distance of a nearby star or a nebula can be accurately measured by the parallax method, the method is ineffective for a far away star. The parallax effect, the shift in the position of a star due to shift in observer's position, is very small. For distances which are too large to measure using parallax, astronomers use

---

[2] Mount Wilson Observatory was founded in 1904 by George Ellery Hale, a pioneer in the field of astrophysics. The observatory is located on top of Mount Wilson (1524 meters) in Pasadena, Los Angeles. It housed two historically important telescopes: the 60-inch Hale telescope, the largest telescope in the world in 1908; and the 100-inch Hooker telescope, world's largest telescope from 1917 to 1949.

[3] A nebula is a cloud in deep space consisting of gas or dirt/dust. The word "nebula" has been derived from the Latin word, meaning "cloud." Earlier, galaxies were also called nebulae due to their fuzzy appearance. Now, a group of numerous stars, dust, planets and other interstellar matter, tied together by gravitational force is known as a galaxy.

"standard candles."[4] Now for a light-emitting object like a star, one defines two types of brightness: intrinsic brightness or luminosity which is the amount of energy radiated by the object per second, and apparent brightness which is the brightness that the object appears to have to the observer. The farther is the observer, the less is the apparent brightness. For a star, d-distance away from Earth, the intrinsic brightness or luminosity, and apparent brightness are related by the simple inverse square law,

$$\text{Apparent brightness} = \frac{\text{Luminosity}}{4\pi d^2}.$$

The relation is easy to understand. We can imagine light rays crossing spherical shells centered on the star (see Figure 5.2). The

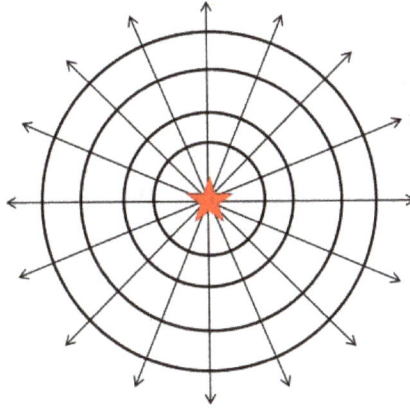

Figure 5.2. Two-dimensional visualization of light rays emanating from a star crossing imaginary spherical shells.

---

[4]A standard candle is a class of astrophysical objects, such as supernovae or Cephid variable stars, which have known luminosity due to some characteristic quality possessed by the entire class of objects.

total energy crossing each shell per second is constant, but since the surface area of a shell at a distance d from the star is $4\pi d^2$, the apparent brightness will be diluted by the same factor.

While the apparent brightness of a star can be measured through a telescope, its intrinsic brightness is not easy to measure. Hubble was lucky that a few years back, American astronomer Henrietta Swan Leavitt (July 4, 1868–December 12, 1921) established a period–luminosity relation for a particular class of stars, called "Cepheid variable" stars. Cepheid variable stars change their luminosity or intrinsic brightness periodically. The periodic variation of a Cepheid variable star is due to the interplay of doubly ionized helium (He^{++}) and singly ionized helium (He^+) ions. He^{++} ions are more opaque than He^+ ions. At the dimmest part of a Cepheid's cycle, the star is enveloped by He^{++}, the radiation cannot escape and the star looks dim. The Cepheid is heated up by the trapped radiation. As the temperature increases, the star expands and cools. As it cools, He^{++} ions in the outer layer capture an electron to become He^+. The radiation can now escape through more transparent He^+, and the star looks brighter. Over time, the expansion stops and reverses due to the star's gravitational attraction. The process then repeats.

Henrietta Swan Leavitt was an American astronomer. In 1892, she graduated from Radcliffe College, Cambridge, Massachusetts. After her graduation, she took a course on astronomy and volunteered to work at the Harvard College Observatory. Sadly, at that time, the scientific profession was almost exclusively a male prerogative; women were not even allowed to operate a telescope. Observatories then used to employ "human computer" whose job was to catalogue star measurements for a measly amount of $0.30 an hour. Leavitt was appointed as a human computer at the Harvard College Observatory. Edward Charles Pickering, director of the observatory, assigned her to catalogue Cepheid variable stars. In the course of her work, Henrietta noticed that intrinsic luminosity or brightness of the Cepheid variable stars and their periods show a definite relationship — the brighter the star, the longer is the period.

Her results, which is known as "period–luminosity" relationship, paved the way for future observational astronomy; a Cepheid variable stars can be used as a standard candle. Unfortunately, Henrietta's boss, Edward Charles Pickering, continued to ignore Henrietta and ultimately, she died of cancer, largely unknown and unrecognized.

Using Henrietta's method, Hubble measured the distances of a large number of nebulae. He found several nebulae at a distance far away from the Milky Way, irrefutably negating the earlier propositions that entire Universe consists of the Milky Way. Indeed, he found the Andromeda galaxy, our closest neighbour, at a distance of 900,000 light years[5] away. Hubble published his results, first in *New York Times* (November 23, 1924), and then more formally at the meeting of the American Astronomical Society on January 1, 1925. The result itself was magnificent, profoundly changing our notion of the Universe. However, Hubble did much more. He measured the velocities of the galaxies.

In 1842, Christian Doppler (November 29, 1803–March 17, 1853), an Austrian physicist discovered an effect which many of us have experienced. When a car blaring its horn approaches us, the sound appears to be higher-pitched than when it is receding from us. The difference is now understood as due to the Doppler effect, named after its discoverer. Christian Doppler was born in Salzburg, Austria. Son of a successful stonemason, he was physically weak and could not join his father's business. He studied philosophy in Salzburg, and mathematics and physics at the Vienna University of Technology and the University of Vienna and graduated in 1825. For several years, he had to struggle for a suitable employment. At one time, he even thought of migrating to the United States of America. Ultimately, in 1835, he obtained a teaching position at a technical secondary school in Prague. Six years later, he was

---

[5] Astronomers measure stellar distances in unit of light years. Light travels at a speed of 186,000 miles per second. The distance traversed by light in one year is called a light year.

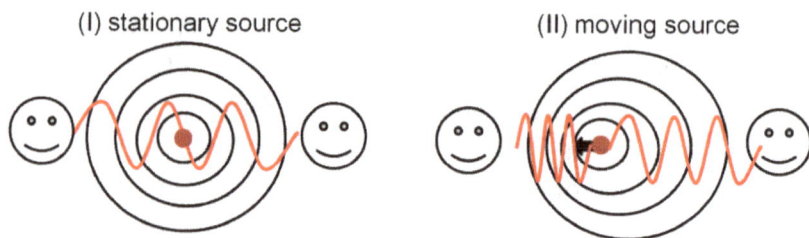

Figure 5.3. Pictorial depiction of the Doppler effect. In (I) the source is stationary and the observers receive sound of the same wavelength or frequency as emitted by the source. In (II) the source is moving in the left direction. The right observer receives sound of the longer wavelength or lesser frequency because the source is receding from him; the left observer receives sound of the lesser wavelength or higher frequency because the source is approaching him.

appointed a professor at Prague Polytechnic. In 1842, he published his most important work, "On the colored light of the binary stars and some other stars of the heavens," where he proposed the change in frequency of a wave for an observer moving relative to its source. The physics of the effect is illustrated in Figure 5.3. If the light source (star) is moving away from us, then due to the Doppler effect, the wavelength of the emitted light is increased or the frequency is decreased. The effect is called red shift (the wavelengths are increased towards red). Lights will be blue shifted if the star moves towards us.

The red shift of a star can be measured by absorption spectrometry. As we know, visible light is composed of many colors; each color has a range of frequency or wavelength. Visible light, if passes through a prism decomposes into a continuous spectrum of colors. It is schematically shown in Figure 5.4. Now if the light is allowed to pass through a gas, some wavelengths, characteristic of the gas, are absorbed and dark lines are seen in those places. The dark lines are called absorption lines. Now if the source is moving away, then due to the Doppler effect, the absorption lines will be red shifted.

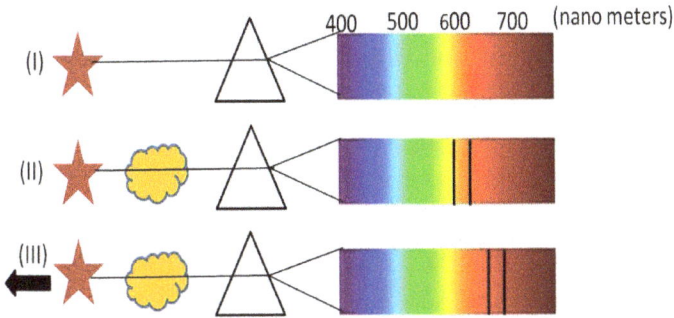

Figure 5.4. A schematic depiction of the absorption spectroscopy. (I) Visible light if passes through a prism decomposes into a continuous spectrum, (II) dark lines in the spectrum if the light passes through a gas, (III) dark lines are shifted towards red if the source moves away.

If λ_0 is the missing wavelength from the source at rest and λ is the missing wavelength from the moving source, the red shift z can be calculated as,

$$z = \frac{\lambda - \lambda_0}{\lambda_0}.$$

Like any hot objects, stars emit radiation called blackbody radiation. The radiation has to pass through the outer gaseous layer of the star. In the process, some wavelengths or frequencies, characteristic of the gas or gases in the outer layer, will be absorbed, producing black lines in the spectrum. If the medium through which the light has passed is identified from the missing wavelengths, one can then easily measure the red shift. From the measured red shift, one can then calculate the velocity of the star. The two are related simply,

Velocity of star = red shift × velocity of light.

Using the Hooker telescope, Hubble measured the red shift of distant nebulae and determined their velocities. The result astonished him. Velocities of the galaxies are in direct proportions to their distances, the further away is the galaxy the more is its velocity. The proportionality constant is now called the Hubble constant. Strictly speaking, it is not a constant with time, but at any instant, it is a constant for all the galaxies. Incidentally, Hubble was not the first astronomer to measure velocities of distance galaxies. Before him, in 1922, American astronomer, Vesto Melvin Slipher measured velocities of 41 galaxies. He also found that most of the galaxies showed positive red shift, i.e. they were receding from the Earth. It seems that our position as an observer is repulsive to the rest of the Universe. This does not mean that we are at the center of the Universe. The Universe has no center. No matter which galaxy you are in, all other galaxies are receding from you. Hubble's results can be understood in an expanding Universe. If the Universe expands with time, the distances between the galaxies will be increasing. Hubble's discovery of the expanding Universe, which was published in 1929, also solved an old puzzle, "why is the night sky black?" The question was asked by the German astronomer Hendrik Wilhelm Olber and came to be known as Olber's paradox. If the Universe is static, and uniformly filled with stars, then over time, light from the stars should reach the Earth and the night sky should have been illuminated; it will be as bright as the day. On the contrary, the night sky is black. The puzzle is solved now. In an expanding Universe, in a finite time period, the lights from the far away stars cannot reach the Earth.

5.2 We started with a Big Bang

Hubble's discovery of the expanding Universe also led to the currently accepted theory, the "Hot Big Bang model" for the evolution of the Universe. Imagine going backward in time. As we go back in time, the Universe becomes smaller and smaller in size. One can conceive a time when the entire Universe was contained in a point, an initial singularity. Prior to that time, there was nothing, the Universe started from the singularity (which scientists calculated to have occurred 13.7 billion years ago). However, size is not the appropriate parameter in this context. For example, if the Universe is presently infinite in size, then, back in time, it was also infinite. We can talk more sensibly in terms of the density, e.g. number of galaxies per unit volume. Hubble's expanding Universe demands that back in time the Universe was denser and hotter. Eventually, one can consider a time when the Universe was infinitely dense and infinitely hot, the singular point. In the Hot Big Bang theory, from the singular point, the Universe started to expand, creating space-time along with. The last sentence needs clarification. Before the beginning of the Universe, space-time did not exist. Space-time was created along with the Universe as it expanded. Thus the Big Bang or expansion of Universe was unlike a bomb explosion; rather it was like the expansion of a balloon. The surface of the balloon is stretched out as we blow it. Imagine that the galaxies are on the surface of the balloon. As the balloon expands the separation between the galaxies increases. A cartoon of the above scenario is shown in Figure 5.5 (for a large sized balloon, a segment of the surface can be assumed to be flat).

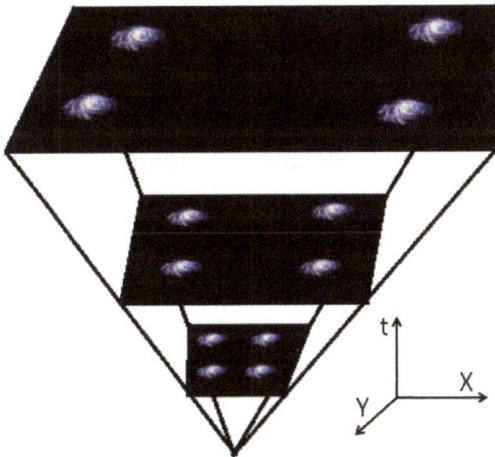

Figure 5.5. Schematic diagram of the Hubble expansion of Universe.

Incidentally, the term Big Bang was coined by one of the fierce opponents of the model, Sir Fred Hoyle (June 24, 1915–August 20, 2001), British mathematician and astronomer. After his education at Cambridge, Hoyle joined the British navy, and spent six years during World War II working on radar development. There he made friendship with Hermann Bondi and Thomas Gold. In war time the three friends used to discuss problems of cosmology. The friendship continued even after the war. In 1945, at the end of war, Hoyle returned to Cambridge as a lecturer in mathematics. He continued to teach at Cambridge till 1972. Hoyle strongly opposed Hubble's expanding Universe model. He did not believe that the Universe had a beginning. He thought that it was a disguised argument for a creator of Universe (otherwise, who would decide when to begin). In one of his BBC interviews, he said:

> "The reason why scientists like the 'Big Bang' is because they are overshadowed by the Book of Genesis. It is deep within the psyche of most scientists to believe in the first page of Genesis."

With his friends, Bondi and Gold, Hoyle formulated an alternate theory called steady state theory of Universe. In his theory, the Universe is eternal, there is no beginning nor end. With time, as the Universe expands, it continually creates matter such that the "cosmological principle," i.e. the Universe is spatially homogenous and isotropic, is maintained. Steady state theory could explain all the experimental facts known at that time and was quite popular in 1950–60s. In a 1949 BBC interview, Hoyle speaking on Hubble's expanding Universe model pejoratively used the term "Big Bang." The term stuck.

Hubble's discovery of expanding Universe has profoundly changed our understanding of the Universe. It is regarded as one of the important milestones in science. Though Hubble is credited with the discovery, two years before Hubble, a Belgian priest and astronomer, George Henri Joseph Édouard Lemaître (July 17, 1894–June

20, 1966) discovered it. George Lemaître was interested both in science and theology. After a classical education in a Jesuit school, he enrolled in the Belgian University, Université Catholique de Louvain, for the civil engineering degree. His education was interrupted by World War I. He served the Belgian army as an artillery officer. After the war, he studied physics and mathematics and obtained his Ph.D. degree. In 1923, Lemaître was ordained to priesthood. In the following year, he moved to Cambridge, England, to work with Sir Arthur Eddington. Eddington initiated him to modern cosmology. He spent some time also at the Massachusetts Institute of Technology, USA. In 1925, he returned to Belgium and joined Université Catholique de Louvain as a lecturer, later became a full professor and continued his life there.

In 1927, Lemaître published a remarkable paper (in French), in a relatively obscure Belgian journal, *Annals of the Scientific Society of Brussel.* The English translation of the paper's title runs as, "A Homogeneous Universe of Constant Mass and Increasing Radius Accounting for the Radial Velocity of Extra-Galactic Nebulae." In the paper, Lemaître solved Einstein's equations for a dynamic (or time-dependent) Universe. Einstein himself solved for a static Universe. Solving Einstein's equations Lemaître deduced that the Universe is expanding. Indeed, Einstein's equations, together with the cosmological principle, always lead to an expanding Universe. Lemaître was not aware, but in 1922, the Russian physicist, Alexander Alexandrovich Friedmann (June 29, 1888–September 16, 1925), also solved Einstein's equations and deduced that the Universe is expanding. However, Lemaître went beyond mere theoretical calculations. He used Vesto Slipher's measurements to establish his results and determined the rate of expansion (what nowadays is known as the Hubble constant). Lemaître also anticipated the Big Bang theory. He argued that far enough back in time, the entire Universe was in an extremely compact and compressed state, forming what he called the "primal atom" and some instability caused by radioactive decay of the primal atom could trigger an immense explosion initiating the expansion. Unfortunately, Lemaître's paper in the obscure Belgian journal went largely unnoticed. Much later, an English translation was published, omitting some vital paragraphs of the original paper. However, the paper did not escape the attention of Einstein. He did not believe it and reportedly, commented to Lemâitre, *"Your calculations are correct, but your grasp of physics is abominable."*

5.3 Nucleosynthesis: How the elements are formed

While Lemâitre and Friedmann laid down the seed of the Big Bang model, it was given respectability by the Russian scientist George Gamow (March 4, 1904–August 19, 1968). Gamow was one of the few colorful personalities in science. As noted by Edward Teller, colloquially known as the "father of hydrogen bomb,"

> "Gamow liked to live, liked fun and liked to make other people have fun."

George Gamow perhaps is more widely known as a popular science writer than as a scientist. He raised the popular science writing to a fine art. His two books, *Mr. Tompkins in Wonderland* and *Mr. Tompkins Explores the Atom*, probably introduced millions of readers to the concepts of relativity and atomic and nuclear physics.

On his 13th birthday, Gamow's father presented him a small telescope with which he patiently studied the stars and decided to become a scientist. He was educated at the University of Leningrad. His fame grew when in 1928 he solved the problem of alpha decay. In alpha decay, some nucleus emits an alpha particle (or helium nucleus), e.g. polonium spontaneously emits an alpha particle and converts into a lead nucleus. Classically such emissions are not possible as the emitted alpha particle has to cross a Coulomb barrier created by Coulomb repulsion between the alpha particle and the rest of the protons in the nucleus. The Coulomb barrier can be understood as follows: imagine an alpha particle approaching a lead nucleus from infinity. As it comes closer, the two protons in the alpha particle will be repelled by the protons in the lead. Only if that repulsion is overcome, the alpha particle can fuse with the lead to make a polonium nucleus. In other words, an alpha particle has to cross a Coulomb barrier to fuse. Similarly, if an alpha particle tries to break out from a polonium nucleus, the barrier has to be crossed again. Gamow proposed a quantum mechanical process whereby the alpha particle can tunnel through the barrier. It was the first application of quantum mechanics in nuclear physics (before it, quantum mechanics was used only to explain atomic phenomena and it was uncertain whether quantum laws were valid in other domains also).

Gamow worked in various European laboratories till 1931, when he was called back to Russia. However, free-natured Gamow could not endure the oppressive atmosphere of the Stalin-ruled Russia. He and his wife made several official attempts to escape from Russia but the Russian government continued to deny them visa. He also made several unofficial attempts to leave Russia but failed. In 1933, Gamow and his wife were issued visa to attend the 7th Solvay Conference.[6] After the conference, Gamow did not return to Russia but continued in various laboratories in Europe and finally settled in the US.

Gamow was interested to explain the abundances of chemical elements. Of all the elements in the Universe, approximately 25% is helium and 75% is hydrogen, with other elements in negligible fractions. He reasoned that the relative abundances of chemical elements were determined by physical conditions existing in the

---

[6] The Solvay conferences were initiated by Ernest Solvay, a Belgian chemist and industrialist. The first in 1911 and the fifth in 1927 were particularly noteworthy, as they helped to define the foundations for the old and new quantum theory. The 1927 conference was also noted for famous the Einstein–Bohr debate on the meaning of quantum nature.

Universe during the early stages of its expansion. In the expanding Universe scenario, one can imagine a time when the Universe consisted only of hot, dense plasma of protons, neutrons, electrons, neutrinos, and photons. It can be understood as follow: imagine oxygen gas being heated. In oxygen atom, 8 electrons revolve around a nucleus consisting of 8 protons and 8 neutrons. The electrons do not fly off due to the attractive electromagnetic interaction between protons and electrons. However, if the gas is heated sufficiently, the electrons can overcome the attractive electromagnetic force and fly off, their thermal energy become so large that they can no longer be bound to the oxygen nucleus by the electromagnetic interaction. The matter then consists of free electrons and positively charged oxygen nuclei. In oxygen nucleus, protons and neutrons are held together by the strong nuclear force. If still heated, the thermal energy will exceed the nuclear force and protons and neutrons will be free, no longer bound in a nucleus. Now the free neutron is not a stable particle. In the time scale of 10 minutes, a neutron decays into a proton, an electron and an antineutrino. Very early in time, our Universe was hot and dense, consisting of protons, neutrons, electrons and neutrinos. As the system expands and cools, a neutron can capture a proton to form a deuteron. A chain of fusion reactions can be established to form the low mass nuclei. Later, as the Universe continues to expand and cool, a stage will come when the positively charged nuclei capture electrons to form atoms. Gamow, with his student Ralph Alpher, calculated the abundances of hydrogen and helium in the Big Bang model of nucleosynthesis. The calculated abundances matched the experimental helium and hydrogen abundances, validating the model. In 1948, the results were published in what is now known as Alpha, Beta and Gamma paper. Ralph Alpher, Hans Bethe and George Gamow, were named as authors of the paper. Incidentally, Bethe was not involved at any stage of the work. Gamow, with his quirky sense of humor, added the name of Hans

Bethe, such that the authors' names resonance with alpha, beta and gamma, the three Greek letters.

We have noted earlier that a nucleus heavier than iron cannot be produced in the fusion process. How are they produced? They are produced in supernova explosion by processes called rapid neutron capture (r-process) or slow neutron capture (s-process) or by proton capture (sp-process). In r-process a nucleus can capture in quick succession a number of neutrons to form an unstable nucleus which then decays to a stable nucleus by emitting an electron and neutrino. In stellar environment, if a nucleus X with atomic number Z and mass number A, captures a neutron, it becomes a nucleus with mass number $A + 1$ while the atomic number remain unchanged:

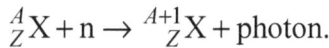

$$_Z^A X + n \rightarrow {}_Z^{A+1} X + \text{photon}.$$

In general the resulting nucleus is unstable and it becomes stable by emitting an electron and antineutrino,

$$_Z^{A+1} X \rightarrow {}_{Z+1}^{A+1} Y + \text{electron} + \text{antineutrino}.$$

This is the slow neutron capture process whereby one obtains a nucleus with one proton number higher. In the rapid neutron capture process, if neutron concentration is high then before the unstable nucleus decays, it can capture several neutrons in quick succession producing a higher atomic number and mass number nucleus:

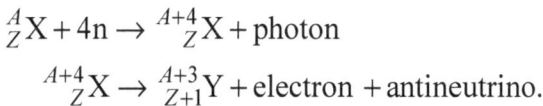

$$_Z^A X + 4n \rightarrow {}_Z^{A+4} X + \text{photon}$$

$$_Z^{A+4} X \rightarrow {}_{Z+1}^{A+3} Y + \text{electron} + \text{antineutrino}.$$

In proton capture processes a nucleus captures a proton to go to a higher mass element:

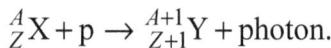

$$_Z^A X + p \rightarrow {}_{Z+1}^{A+1} Y + \text{photon}.$$

It is comparatively rare as charged protons have to overcome the Coulomb barrier and can occur only in very high temperature environment.

5.4 CMBR, the relic of Big Bang

The Universe, beginning with a Big Bang, also left another relic, the cosmic microwave background radiation (CMBR). Scientists' confidence on the Big Bang theory grew after detecting the relic radiation. Origin of cosmic microwave background radiation can be understood as follows.

Prior to the neutral atom formation, the Universe consisted of hot plasma of electrons, baryons (protons and neutrons) and photons. They were constantly interacting with each other. As the system expanded and cooled, neutrons and protons combined to form nuclei. Later on, as the Universe cooled to around 3000 kelvins, electrons got attached with the nuclei to form atoms. At this stage, energy of the photons was too low to interact with the atoms, the photons got decoupled from the system and began to travel freely through space. Those photons can be observed today, considerably cooled due to expansion of the Universe. In 1948, Ralph Alpher and his coworker Robert Herman predicted the existence of relic, blackbody radiation at temperature of 5 kelvins, pervading the entire Universe uniformly. However, they could not convince fellow experimentalists to use the characteristic blackbody form of the radiation to find it. The radiation was finally detected in 1964, accidentally, by two scientists, Arno Penzias and Robert Woodrow Wilson, at the Bell Telephone Laboratory. They were building a sensitive microwave horn antenna.[7] One of the oldest horn antennas was built by the Indian scientist Sir Jagadish Chandra Bose, to demonstrate wireless transmission. Those antennas are called horn

---

[7]An antenna is an electrical device that receives or transmits electromagnetic waves.

antenna as the wave guide[8] has a flaring, horn-like structure. Penzias and Wilson were disturbed by the presence of a noise, corresponding to radiation at 3 kelvins. After many attempts to get rid of the noise, they decided that it was a genuine signal. Unknowingly, they have discovered the relic of Big Bang. They were awarded the Nobel prize for their discovery. Unfortunately, Alpher and Herman, who predicted the signal were not considered for the prize. In later years, the relic radiation was measured in various experiments more and more accurately. In 1989 NASA launched a satellite called Cosmic Background Explorer (COBE) and in 2001 the Wilkinson Microwave Anisotropy Probe (WMAP) to measure the relic radiation and its anisotropy — its dependence on direction. The relic radiation was found to be remarkably uniform over the entire sky, at an effective temperature of 2.73 kelvins. The fluctuations in its temperature are tiny, only one part in 100,000.

The extremely uniform CMBR temperature poses a problem: why is our Universe isotropic to such a great extent? The only way for two regions to have the same temperature is that they are close enough to each other for information to be exchanged between them so that they can equilibrate to a common temperature. The fastest speed that information can travel is the speed of light. If two regions are far apart such that light has not enough time to travel between them, the regions are isolated from each other. There is then no reason that those two regions will have a common temperature. A precisely isotropic cosmic microwave background radiation however, appears to defy the law that information cannot travel faster than light. If we look to the west we detect cosmic background radiation at 2.73 kelvins. We look to the east we detect cosmic background radiation, at exactly the same temperature,

---

[8] In open space, an electromagnetic wave like microwave propagates in all directions. A waveguide consists of a rectangular or cylindrical metal tube or pipe that guides the wave to move in a particular direction with minimum loss of energy. This is achieved by total reflection from the surface of the wave guide.

2.73 kelvins. Yet, the radiation from the east and the radiation from the west are separated by approximately by 28 billion light years. This problem is called the horizon problem. The horizon problem was finally explained by introducing a concept called "inflation." Briefly, immediately after the Big Bang, our Universe with intimately connected matter and radiation underwent an exponential expansion such that its size increased by a factor of 10^{50}! The information contained in the pre-inflationary Universe did not have to travel at the speed of light; it traveled at the speed of inflation.

Early inflation also solved another serious problem for the cosmologists. The problem is called the flatness problem: why is our Universe flat? In Einstein's theory of general relativity, gravity determines the space-time curvature. In other words, space-time curvature is affected by the presence of matter and radiation. Depending on the density of matter and radiation, one can have three types of Universe: (i) open Universe, density of Universe is less than a critical density; (ii) close Universe, density of Universe is more than the critical density; and (iii) flat Universe, density of Universe is equal to the critical density. Illustrations of three types of Universe are shown in Figure 5.6.

The fluctuations in cosmic microwave background radiation can be correlated with density of the Universe and it turned out that presently, within 1%, the density of our Universe is just the right amount for a flat Universe. Now Universe has expanded for about 13.7 billion years. The 1% difference then implies that at the beginning, the density of Universe was fine tuned to the level of 10^{-62} for a flat Universe. Such a level of fine tuning is hard to accept. In the inflationary model, a Universe with an arbitrary density will expand exponentially to take the density arbitrarily close to the critical density. Subsequent evolution of the Universe will cause the value to grow, bringing it to the currently observed value of around 0.01.

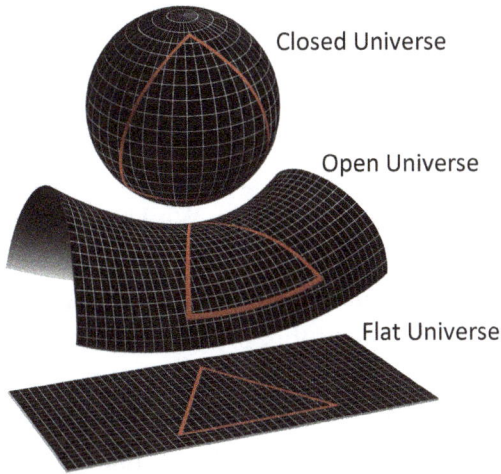

Figure 5.6. Local geometries of a closed Universe, an open Universe and a flat Universe are shown. Image courtesy of NASA.

5.5 Universe is mostly invisible

Hubble's discovery of expanding Universe startled the world. However, the Universe had more secrets in store: it had dark matter — the matter that does not emit any kind of electromagnetic radiation and is invisible to us. In fact, the secret was revealed in 1933 by the Swiss astronomer, Fritz Zwicky (February 14, 1898–February 8, 1974), he even coined the term "dark matter," but his contemporaries chose to ignore him. Son of a Swiss industrialist, Fritz's family wanted him to study commerce. However, in the course of time, his interest shifted to mathematics and physics. In 1922, he obtained his doctorate with a dissertation on ionic crystals. Three years later, on an international scholarship from the Rockefeller Foundation, he moved to the California Institute of Technology to work with the famous physicist Robert Millikan. Millikan once said that he kept Zwicky around because some of his hair-brained ideas might turn out to be right — which many of them did! Two years after Chadwick's discovery of neutron, Zwicky speculated about neutron star — an ultra-compact star with a core of neutrons; he also initiated the general morphological studies — a scientific enquiry where all the parameters that might be of importance are first defined and their interrelationship is investigated; posited that galaxies could act as gravitational lenses; suggested supernova as distant indicator etc. While he did not get much recognition during his lifetime, today, when researchers talk about neutron stars, dark matter, and gravitational lenses, they all start the same way: "Zwicky noticed this problem in the 1930s. Back then, nobody listened ..."

In 1933, Zwicky was measuring the red shift of some distant galaxies. He was also interested to know the mass of those galaxies. Now the mass of a galaxy and its luminosity is related. The more massive is the star, the more luminous it is. It can be understood. A star has to burn its fuel to balance against the gravitational attraction trying to collapse it. The more massive is the star, the stronger is its gravitational pull and to counter the pull, the star has to burn its fuel at an increased rate (more massive stars then die sooner). The relation between mass and luminosity is not a linear one, rather luminosity increases as some power (between 3 to 4) of the mass. Fritz wanted to obtain the galaxy mass in an alternate manner. From the observed red shift, he found the velocities of galaxies. Then he applied the virial theorem. The virial theorem states that for a stable, self-gravitating, spherical distribution of equal mass objects (stars, galaxies, etc.), twice the total kinetic energy[9] of the system is equal to its total gravitational potential energy.[10] What Zwicky found puzzled him: the mass of galaxies from virial theorem greatly (approximately by a factor of 10) exceeded that from luminosity measurements. It appeared that most of the galaxy masses did not emit light. He coined the term "dark matter," the matter which does not emit any light or

---

[9] The kinetic energy of a body is the energy associated with its motion.

[10] The potential energy is the energy stored in an object by virtue of its position.

electromagnetic signal, i.e. invisible to us. However, Zwicky's result did not attract much attention.

Dark matter came into forefront only in 1970s when American astronomer Vera Rubin obtained some irrefutable evidence for its existence. Born in 1928, Vera Rubin had her early education in New York. For her master's, Vera wanted to enroll at Princeton, but Princeton denied her admission. Princeton did not allow women in the astronomy program until 1975. She obtained her master's from Cornell University. Under the guidance of George Gamow, she wrote a controversial dissertation and later joined Georgetown University as a faculty. Vera was the first woman allowed to operate a telescope at the Palomar Observatory. Prior to her, women were not allowed to operate telescopes. In general, galaxies are classified into three types (see Figure 5.7): (i) spiral galaxy, shaped like a flattened disc with one or more arms; (ii) elliptical galaxy, round or elliptical in shape; and (iii) irregular galaxy, neither spiral nor elliptical. Our own galaxy is a spiral galaxy. While trying to understand the variation of brightness in spiral galaxies, Vera stumbled upon the dark matter. She and her collaborator, Kent Ford (who designed a sensitive spectrometer[11]) measured the galaxy velocity as a function

spiral galaxy	elliptical galaxy	irregular galaxy

Figure 5.7. The major types of galaxies are shown. Image credit goes to NASA.

---

[11]A spectrometer is an instrument used to measure properties of light over a specific portion of the electromagnetic spectrum. We know that when (white) light passes through a prism, it breaks into different color components. Crudely speaking, a spectrometer is a much improved version of a prism.

of distance from the center. Since most of the galactic mass is concentrated in the central bulge of the galaxy, one expects that galaxy
velocity will decrease with distance. To her surprise, the velocity did
not decrease with distance. The nearly constant velocity as a function of the radial distance indicated the presence of non-luminous or
dark matter. She made detailed measurements and produced irrefutable evidence that the observed galactic velocities required the
existence of dark matter.

Over the years, many more measurements were made and it was
proved that in our Universe, only 5% of matter is visible, approximately 25% of matter is dark and the rest is dark energy (to be
explained later). Even though we know there is dark matter in abun-
dance, it is yet to be observed directly. Dark matter is not easy to
detect, it can be detected only by the gravitational effect it produces,
e.g. by gravitational lensing effect. We also do not know what it is
made of. Currently, there are several theoretical speculations:
weakly interacting massive particles or WIMPS, massive astrophysical compact halo object or MACHO, a hypothetical particle
called axion, the lightest supersymmetric particle called neutralino
etc. Needless to say, these hypothetical particles are yet to be
detected and presently, we do not know whether or not any of them
constitutes the dark matter.

5.6 Oh! Still accelerating

Hubble discovered that the Universe is expanding; the galaxies are
receding from each other. Now gravity is attractive. Thus, one
expects that eventually, the Universe will slow down. Indeed, in a
simple mathematical model of cosmology, the expansion history
depends only on the mass density of the Universe. The greater the
mass density, the more the expansion is slowed down by gravity.
Unarguably, mass density was higher in the past and the Universe
would have been expanding much faster than it is today. Scientists

wanted to verify the slowing down. They reasoned that light from a
star one light year away observed now had started one year back.
Light from a star ten light years away started ten years earlier. Then
if we can measure light from very-large-distance stars, we can know
about the expansion at a very early time.

By Hubble's law, distance of a star is proportional to its velocity
and velocity in turn is proportional to the red shift. The stars at very
large distance will have high red shift (z). Astronomers tried to
measure large-distance stars by looking for high red shift. However,
very-large-distance stars are so faint that they cannot be measured.
Way back then, Fritz Zwicky and his collaborator Walter Baade
suggested that supernovae could be used as a distance indicator.
Supernova explosions are very bright and over a few weeks, out-
shine an entire galaxy. Supernovae are classified into two types,
type I and type II. Type II supernovae form from the core-collapse
of a massive star. They display prominent lines of hydrogen in their
spectra and have irregular light curves.[12] Type I supernovae form
from the explosion of a white dwarf star. In a close binary system,
a white dwarf can gain mass and explode as supernova. Type I
supernovae have no hydrogen lines in their spectrum. The classifi-
cation of type I supernovae can be further divided by using the
helium and silicon lines in its spectrum: type Ia, presence of silicon
line and absence of helium line; type Ib, presence of helium line and
absence of silicon line; and type Ic, absence of helium and silicon
line. Type Ia supernovae, characterized by the absence of hydrogen
line and presence of silicon line, are ideal for distance measurements. Wherever they occur, supernovae Ia produce a precise standard of brightness. However, type Ia supernova explosions are rare,

---

[12] Light curves are graphs that show the brightness of an object over a period of
time. Astronomers generally know the light curves from standard objects. They
can compare the light curve from a new source to those standard light curves and
identify the source.

approximately one in 500 years in a galaxy. They are also random and one does not know where to look for. Above all, they are short lived — they brighten and fade away in the time scales of weeks.

Given the rarity (one in 500 years in a galaxy) and randomness, it appeared to be impossible that one in his lifetime could use type Ia supernova as a standard candle, and measure the distance of a galaxy. However, the impossible was made possible by the ingenuity of the American astrophysicist Saul Perlmutter. Saul Perlmutter was born on September 22, 1959, in an educated Jewish family. He graduated from Harvard *magna cum laude*[13] in 1981 and received his Ph.D. from University of California, Berkeley in 1986. He joined the UC Berkeley Physics Department in 2004. In 1988 Perlmutter initiated the Supernovae Cosmology Project (SCP) to measure the presumed deceleration of the Universe — using supernovae Ia as standard candles. Perlmutter and his collaborators devised a method called "supernovae on demand." The method is schematically explained in Figure 5.8. For two to three nights, just after the new moon, with a 4-m telescope coupled with a CCD (charge coupled device)-based imager, the group would observe a

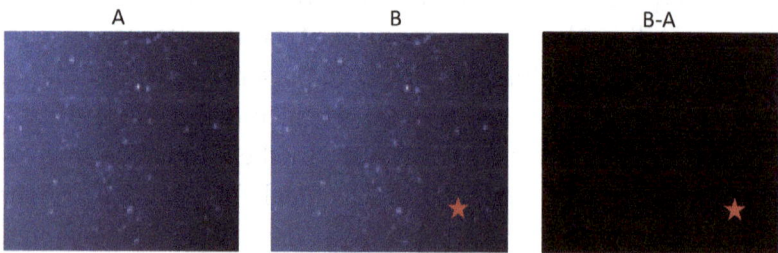

Figure 5.8. Principle of supernova on demand. (A) shows the photograph of a patch of sky. The same patch is photographed three weeks later (B). The subtracted image (B-A) reveals a new source of light.

---

[13] A type of honours degree in Harvard.

patch of sky with thousands of galaxies. The same patch of the sky would be observed again after three weeks. With advanced image processing techniques, the two images could be processed to reveal supernovae. If a candidate supernova is detected, follow up observations were arranged with the world's largest telescopes in Chile, Hawaii and La Palma.

The method was a success and the first high-z supernova was discovered in 1992; and by 1994, the total number found by SCP reached seven. In 1994, a competing collaboration, High-z Supernovae Search Team (HZT) was initiated by American astrophysicist Brian Schmidt. Schmidt was born on February 24, 1967, in the American state of Montana. Schmidt had his schooling at Alaska. Family members of Vesto Melvin Slipher, who in 1922 first measured the distant galaxies, instituted a scholarship for the worthy students to pursue science at the University of Arizona. Schimdt's undergraduate education at the Arizona University was funded by the Vesto Melvin Slipher scholarship. After his Ph.D. from Harvard University, he joined the Australian National University (ANU). American astrophysicist Adam Guy Reiss (born December 16, 1969) also joined Schmidt's team. In 1992, Reiss graduated from the Massachusetts Institute of Technology. After his graduation, he joined the Astrophysics Department of Harvard University and in 1994 received his master's degree and in 1996 his doctoral degree. Following his Ph.D. from Harvard University, in 1996 he joined University of California, Berkeley, as Miller Research Fellow. Schmidt's HZT team, independently of Perlmutter's team, searched for high-z supernovae.

When Saul Perlmutter or Brian Schmidt initiated their supernova search program, they could think of only two possible scenarios for the fate of the Universe (see Figure 5.9): (i) if the Universe started with a lot of material it would be expanding much faster in the past, it would also be slowed down a lot, eventually coming to a halt and then start to contract. It would end in gnaB

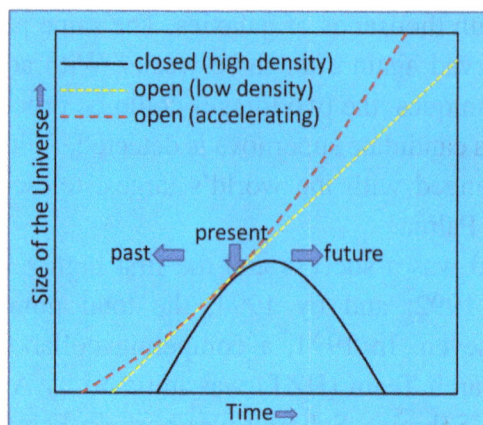

Figure 5.9. Possible expansion history of the Universe. Supernova search experiments, SCP and HZT conclusively proved that our Universe is accelerating.

giB[14] (that is the Big Bang written backwards — Big Crunch), (ii) If there weren't enough material, the Universe would be expanding at about the same speed in the past as now, and would continue to expand forever. Both groups were perplexed by their findings. Their experimental results did not fit either of the two scenarios, rather threw up a hitherto unconceivable situation — the Universe is accelerating. In 1998 and 1999, the two groups published their results, HZT based on 16 supernovae and SCP based on 42 supernovae measurements. Based on their measurements, both groups concluded that the expansion of the Universe did not slow down but actually accelerated. The Universe is not only expanding, it is accelerating! The discovery led to the Nobel Prize to Perlmutter, Schmidt and Adam Riess. An artist's depiction of evolution of Universe, since its creation in the Big Bang, is shown in the beautiful Figure 5.10.

---

[14]The term "gnaB giB" first appeared in a popular science book, *The Restautant at the End of the Universe* by Douglas Adams:

> "But what about the End of the Universe? We'll miss the big moment."
> "I've seen it. It's rubbish," said Zaphod, "nothing but a gnab gib."
> "A what?"
> "Opposite of a big bang. Come on, let's get zappy."

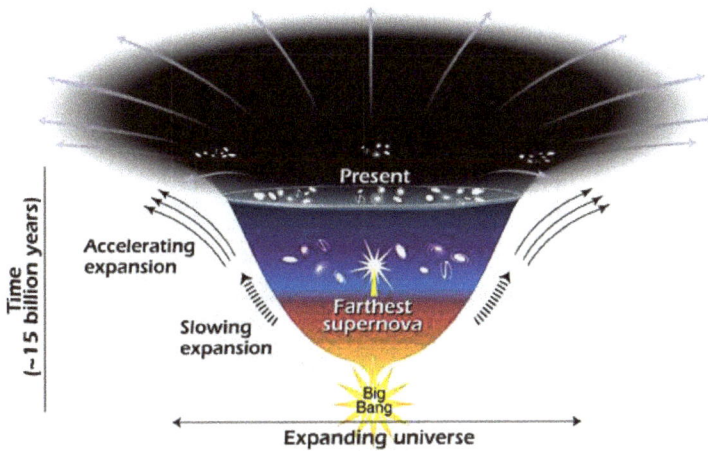

Figure 5.10. An artist's depiction of evolution of the Universe, since its creation in the Big Bang. The diagram reveals changes in the rate of expansion since the Universe's birth 14 billion years ago. The more shallow the curve, the faster is the rate of expansion. The curve changes noticeably about 7.5 billion years ago, when objects in the in the Universe began to fly apart at a faster rate. Astronomers theorize that the faster expansion rate is due to a mysterious, dark force that is pushing galaxies apart. Image curtesy of National Aeronautics and Space Administration (NASA) and European Space Agency (ESA).

5.7 Dark energy

The concept of dark energy is crucial in understanding the experimental result that our Universe is accelerating. Why does the Universe accelerate? Gravity always pulls, never pushes out. An accelerating Universe indicates that it contains some matter or equivalently energy, which unlike the ordinary matter or energy is repulsive. It is called "dark energy." From experimental and theoretical studies, it now appears that around 70% of matter or energy of our Universe is repulsive. The matter composition of our Universe, as per our present understanding is shown in Figure 5.11. Visible matter (matter that emits light or electromagnetic radiation) constitutes only 5% of the total matter (or equivalently energy) of our Universe. Of the rest 25% is dark matter and 70% is dark energy.

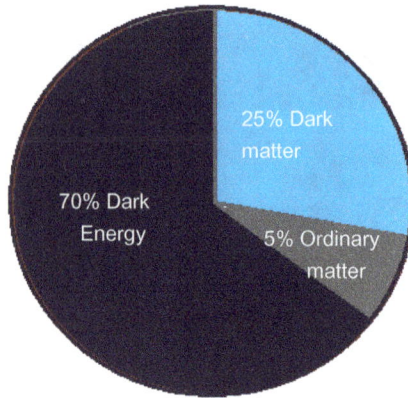

Figure 5.11.   Matter composition of our Universe is shown.

We are still uncertain what this dark energy is. It is one of the greatest mysteries of the present day cosmology. In 1917, Einstein applied his general relativity equations to obtain a dynamic model of the Universe. At that time, the Universe was believed to be static, eternal. Einstein's solutions, on the other hand, indicated that the Universe is expanding. To obtain a static Universe Einstein added a term to his equation which he called "cosmological constant." The cosmological constant acts as a repulsive force and makes the Universe static. However, Einstein's solution was wrong in the sense that the static solution was not stable against any perturbation. Even a minute disturbance would have caused the Universe to collapse. When Hubble discovered that the Universe is expanding, Einstein promptly abandoned the cosmological constant, saying, "it was his greatest mistake." Today, scientists believe that Einstein's greatest mistake might possibly be his greatest achievement. It might be the key to the understanding of dark energy.

The cosmological constant is a property of space-time. One can consider it as the energy associated with the space-time. As the Universe is expanding, it is creating space-time, continually adding energy, which keeps the Universe in accelerating mode. Indeed, in quantum mechanics, due to the uncertainty principle, vacuum is not necessarily empty, rather it is a very active medium, continually creating matter and antimatter in a macroscopically unnoticeable time scale, but whose effect can be quantified and compare with experiments. The cosmological constant could have been identified with the quantum vacuum energy, but for the orders of magnitude difference. The quantum vacuum energy exceeds the cosmological constant by some 120 orders of magnitude.

Another popular theory of dark energy is the "quintessence." The word quintessence originated from the Latin *"quinta essentia"* meaning "fifth essence." Greeks used the word to describe a mysterious fifth element, in addition to "air," "earth," "fire" and "water." It is another name of aether. All the heavenly bodies are made of pure, incorruptible quintessence or aether, it permeates the entire Universe. In cosmological models, quintessence is a scalar field that fills the Universe in addition to matter and radiation, much like Einstein's cosmological constant. However, unlike the cosmological constant, property of quintessence changes with time. Early in the Universe' evolution, it was attractive, but now it is repulsive.

Some scientists also invoked the "anthropic principle" as an explanation of dark energy. In Greek, *anthrop* means man. Over the years, some scientists believed that the Universe was specifically designed to sustain "life." If one examines the nature, one finds that existence of life required some fine tuning of the nature. Take the example of solar light. Unarguably, life would not be possible without solar light; it is our source of energy. Now, Sun is powered by nuclear fusion — two hydrogen ions or protons fuse to form a deuteron nucleus. In the process, a proton is converted into a neutron. It is a slow process. The slow proton to neutron conversion process is responsible for the long life of the Sun. If it was possible for two protons to fuse directly to form a nucleus, then the Sun would have exploded very fast. Now, why two protons do not fuse but one proton and one neutron fuse? It is determined by a very delicate balance between the strong and electromagnetic force. Explicit calculations show that if the strong force was weaker by a few percent, deuteron would not be formed and the Sun would not radiate. If the strong force was a few percent stronger, two protons would fuse and the Sun would radiate very fast and evaporate. In both cases, life would not be there. It is an extraordinary fact that the strength of the nuclear force just happens to lie in the narrow range in which neither of these two catastrophes occurs. This may be considered as an example of the anthropic principle. In another example of fine tuning, consider the case of carbon. Abundant carbon is needed for life. It is the second most abundant element in the human body, after oxygen. How is it formed? Carbon is formed by the so-called triple-alpha process, where three alpha particles (helium nucleus) fuse to form carbon. In triple-alpha process in the first step, two alpha nuclei fuse to form beryllium nucleus which again fuses with an alpha to form the carbon nucleus. But energetic of the process required the nuclear energies to be finely tuned and carbon should exist in a resonant state called Hoyle state (Fred Hoyle, the English astronomer, postulated the excited state even before it was discovered). There are several more such instances of fine tuning that I would not allude to. In 1974, Brandon Carter, an Australian theoretical physicist and astronomer, introduced the notion of anthropic principle. The anthropic principle can be stated as follow:

"The Universe must be such as to admit the creation of observers within it at some stage."

There is a weak version of anthropic principle that can be stated as follow:

"The observed values of all physical and cosmological quantities are not equally probable but they take on values restricted by the requirement that there exist sites where carbon-based life can evolve and by the requirement that the Universe be old enough for it to have already done so."

Simply speaking, by the anthropic principle, the world is the way it is, at least in part, because otherwise there would be no one to ask why it is the way it is. The dark energy is there, because otherwise, we, humans would cease to exist. However, several scientists have objected to the anthropic principle. First of all, the principle does not advance our knowledge about the working of the nature. They believe it is an argument for God in disguise. Instead of saying the Universe is the way it is for the existence of life, we could have very well said the Universe is the way it is by the will of God.

Chapter 6

Looking into the Future

Sometimes I think we're alone in the Universe, and sometimes I think we're not. In either case, the idea is quite staggering.

Arthur C. Clarck

In the preceding pages, I have recounted the evolution of scientific ideas about our Universe. In the beginning, men believed that an omnipotent God has created the Earth, Sun, Moon, and all the heavenly bodies. God also created all the living and non-living beings on the Earth. They believed Earth to be flat shaped, standing on the back of a tortoise (or an elephant); Sun God riding on a chariot traveling around the Earth creating day and night. Their Universe was small, consisting of Earth, Sun, Moon and five planets and fixed stars. Over the ages, men learned about the round shape of the Earth; learned that Earth is not stationary, it is rotating around itself and circling the Sun. We learned about gravity, the force which makes an apple fall on the Earth is the same force that makes the Earth orbits around the Sun. We discovered that Sun was only one of the stars among the billions of stars in a galaxy we named Milky Way, and our Universe contained billions of such galaxies. We learned that the Universe is expanding, which necessitated a beginning for the Universe. We concluded that about 14 billion years ago, the Universe began with a Big Bang. More recently, we also learned that our Universe is not only expanding, it is accelerating.

The concept of Big Bang poses a big question to men. What was before the Big Bang? Why did it take place? Who ordered it? The questions have the similar connotation with the questions asked 20,000 years before, at the beginning of mankind, "who made this Universe? Where did it come from? How and why did it begin?" It may appear that in 20,000 years, we have not made any progress. Knowledge gathered over the years did not make our understanding about Universe any better than it was before. As if we have traveled along a circle and after a lot of labor came back to the point from where we started. Is it really so?

Before we answer the question, it is pertinent to understand what we mean by science or scientific method. As defined in the beginning, science is the intellectual and practical activity encompassing the systematic study of the structure and behavior of the physical and natural world through observation and experiment. How science or scientific method works was best exemplified by Richard Feynman, one of the greatest scientists of the 20th century. In a public lecture, Feynman said:

> "In general, we look for a new law by the following process. First, we guess it. No, don't laugh, that's really true. Then we compute the consequences of the guess, to see what, if this is right, if this law we guess is right, to see what it would imply and then we compare the computation results to nature, or we say compare to experiment or experience, compare it directly with observations to see if it works. If it disagrees with experiment, it's wrong. In that simple statement is the key to science. It doesn't make any difference how beautiful your guess is, it doesn't matter how smart you are who made the guess, or what his name is. If it disagrees with experiment, it's wrong. That's all there is to it."

Guess → Compute Consequences → Compare with experiments → Agree → Theory

Revise guess ← disagree

Compare with experiments → disagree → Revise guess → Compute Consequences

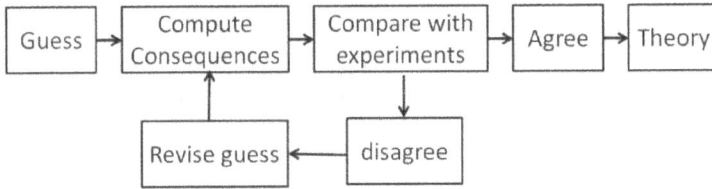

Figure 6.1. Flow chart showing the basis of scientific method.

The simple flow chart shown in Figure 6.1 summarily explains the scientific method. You start with a guess and compute the consequences; if consequences of the guess agree with your observations or experiments, it becomes a theory. If they disagree, you go back to the guess, revise it, again compute the consequences and compare with experiments. The process goes on. When a scientist makes his theory public, he generally refrains from mentioning all the trial and error he underwent to arrive at the final theory.

Sir Isaac Newton guessed that heavenly bodies attracted each other and the attraction fell off as the inverse of the square of the distance. He computed the consequences of this guess and found it to be true with experiments. Now, could anybody have made that guess? An emphatic No! A correct guess does not come out of a hat, rather it comes with experience, knowledge, and intuition. It is the intuition that makes a scientist a great scientist. For example, Robert Hook, a contemporary of Newton had the experience and knowledge. However, he lacked Newton's intuition and could not discover the gravitational law. In general, a scientific law is the description of an observed phenomenon. It doesn't explain why the phenomenon exists or what causes it. As an example, consider Newton's second law of motion: *"The alteration of motion is ever proportional to the motive force impressed; and is made in the direction of the right line in which that force is impressed."* The law does not tell us why the alteration of motion is proportional to the motive force. The scientific laws, axioms or postulates are the basis upon which science grows. From there only you can ask, "Why or how?" Moreover, each law has its own domain. For example, Newtonian mechanics is not applicable to the microscopic world. They follow an entirely different mechanics called quantum mechanics.

When you ask, what happened before the Big Bang, you are going out of the purview of the Big Bang model of Universe. Because, in the Big Bang model there was absolutely nothing before the Big Bang, no space-time, nor any law to observe. A scientist cannot answer this question intelligently. The situation may be compared with St. Augustine. Faced with the question, "What God was doing before he created the World?" St. Agustine replied, "He was creating the Hell for those who asked this type of questions."

I have titled the chapter "Looking into the Future." The future is always uncertain. Five hundred years back, nobody would have believed that our solid Earth, with trees and houses, cities and countries, mountains and seas, envelope of air and all, is rushing through the space at the enormous speed of 107,300 km per hour and also at the same time spinning at a speed of 1670 km/hour. Yet, today, we do not even raise our eyebrows. In the following pages, I will venture into our future with the warning to the reader that the future envisaged is solely mine, limited by my understanding and experience, and you have every reason to question it.

In the study of Cosmos, in the past, we have relied upon the Newtonian theory of gravity and later, on Einstein's theory of general relativity. Both theories are classical theories, they are not quantum theories. The primary difference between a classical and quantum theory is that while the former is deterministic, the later is not. A quantum theory is probabilistic in nature. Do we need a quantum theory to study Cosmos? As such quantum theories are for microscopic bodies and inarguably, our Cosmos is macroscopic. However, there is an aesthetical reason for demanding a quantum mechanical description of gravity. Three of the four fundamental forces, electromagnetism, strong and weak force, are amenable to quantum mechanical description. Aesthetic point of view then demands a quantum mechanical description also for the gravitational force. There is an even more demanding requirement. Consider the situation at the Big Bang. The entire Universe is contained in a singularity. In the immediate aftermath of the Big Bang, our Universe is microscopic in scale and is an apt case for quantum modeling. The description of very early Universe does demand a quantum theory of gravity. However, despite several attempts, scientists could not come up with a realistic theory of quantum gravity. Whenever gravity and quantum theory are commingled, an absurd result — infinity — is obtained even for a well-defined process. As such, scientists have faced the problem of infinity in quantum theories. Origin of infinities in a quantum theory can be understood easily. In quantum theory, the vacuum is not necessarily empty, rather it is seething with activity, continuously producing and annihilating particle–antiparticle pairs in a time scale constrained by Heisenberg's uncertainty principle. Any quantum mechanical process, when vacuum activity is taken into account, results into infinity. Scientists have devised a method called renormalization procedure to work with the infinities. The infinities can be arranged such that they are canceled leaving a finite result. Renormalization procedure works surprisingly well for the three forces — electromagnetism, strong and weak forces. For gravity, however, the procedure does not work; quantum gravity is not renormalizable.

A contradiction between quantum theory and general relativity also surfaces when general relativity is applied to the quantum mechanical uncertainty principle. In quantum mechanics, by uncertainty principle, if the position variable of a body is measured with the precision of L, its momentum will be uncertain by: $\Delta p_x \geq \hbar/2\Delta x = \hbar/2L$. In relativity, mass and energy are synonymous, any form of energy E acts as a gravitational mass $M = E/c^2$, where c stands for the speed of light. Now, in general relativity, any mass distorts the space-time; the more mass the more is the distortion of space-time. If mass M is concentrated within a sphere of radius $R \sim GM/c^2$, where G is the gravitational constant, the distortion of space-time is so great that even light cannot come out — a black hole is formed. To an observer beyond the horizon at $R = GM/c^2$, the body is hidden. Unlike in quantum mechanics, where a body can be localized to an arbitrary precision, general relativity then limits the precision to a length scale called Planck length, L_{Planck}:

$$L_{\text{Planck}} = \sqrt{\frac{\hbar G}{c^3}} \approx 1.6 \times 10^{-33}\,\text{cm}.$$

Above is a rather remarkable formula. It involves three fundamental constants: reduced Planck's constant $\hbar = h/2\pi$, speed of light c and gravitational constant G. The three fundamental constants are related to three basic aspects of our Universe: $\hbar$ signifies the quantum aspect, c the relativity and G the gravitational phenomena. The Planck length thus embodies the three basic aspects of our Universe.

Why gravity and the other forces seem to be incompatible? There is a fundamental difference between gravity and the three forces — strong, weak and electromagnetic. The fundamental difference is that while the three forces act on space-time, in general relativity, space-time (or more precisely, its curvature) itself is the gravity. For the three forces, fields are quantized on a space-time continuum, but for gravity, space-time continuum is the force. We need a prescription for quantizing space-time. Presently, two approaches are being perused to quantize gravity. One of the approaches is called loop quantum gravity and the other is called string theory. Both theories are highly complex and require advanced mathematics. In the following, it will be my endeavor to give the reader a limited flavor of these two complex theories.

Earlier, I have mentioned about Faraday's visualization of electric and magnetic field as lines of force and loops. Loop quantum gravity is a quantum mechanical description of gravity in terms of gravitational field analog of Faraday's magnetic field loops, i.e. we visualize gravitational field as self-terminating loops. However, unlike in electromagnetism, where Faraday's lines and loops reside on the background space, there is no background space in loop quantum gravity. It has only loops, interconnected loops forming the space. An example of space in loop quantum gravity is shown in the left panel of Figure 6.2. In loop quantum gravity, space is no longer a continuum that is infinitely divisible, rather it is discrete and has a granular structure, visible only at the Planck scale.

In 1971, British mathematician Roger Penrose, one of the scintillating minds of 20th century, invoked the Machian principle: *"A background space on which physical events unfold should not play a role; only the relationships of objects to each other can have significance."* Penrose introduced a discrete model for three-dimensional continuum space — spin network system. As shown in Figure 6.2, it is a network of line segments and nodes or vertexes formed by three adjoining line segments. Each link is labeled by an

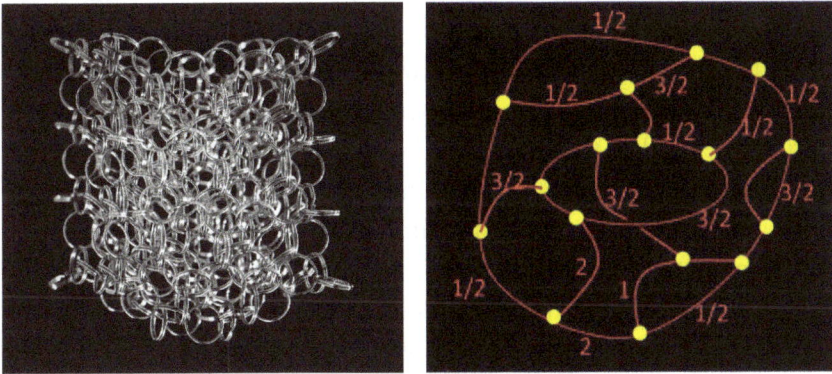

Figure 6.2. Left panel: A net of intersecting loops form the 3-dimensional space in loop quantum gravity. The open source figure is from Wikimedia. Right panel: Space in loop quantum gravity can be replaced by a spin network system. The yellow dots are nodes (intersecting points of loops) and the portion of loop between two nodes is the link. Nodes carry quantum numbers of volume elements and links carry quantum numbers of area elements.

integer or half-integer number called spin quantum number. The spin network system is essentially a diagrammatic representation of the interaction between particles and fields in quantum mechanics and greatly facilitates complicated mathematical calculations.

In loop quantum gravity space is nothing but intersecting loops. Identifying intersecting points with the nodes and the portion of a loop between two nodes with the links, one can have a one-to-one correspondence between the space in loop quantum gravity and the Penrose spin network system. The identification provides an elegant mathematical prescription for quantizing gravity. Einstein's equations for gravity can be written in terms of the loop variables, and quantized. The quantization is in terms of volume. Space is made of tiny volumes (or a number of loops). However, unlike in classical theory, where volumes can have any value, in loop quantum gravity, volumes are quantized, i.e. they can only be an integer multiple of some elementary volume. A volume has a surface. In loop quantum gravity, the surface is also quantized. The volume and surface quantum numbers are carried by the node and link variables. While the complete theory of loop quantum gravity is yet to be formulated, the theory has shown great promise. For example, the theory reproduces the famous Bekenstein–Hawking area law for the black hole entropy. Another interesting consequence of loop quantum gravity is that it gets rid of the initial Big Bang singularity. As such, the Big Bang singularity is not a prediction of general relativity. Rather one can say that the neglect of quantum effect in early Universe led to the singularity. Since volume is quantized in loop quantum gravity, the Big Bang singularity is avoided. Instead of the Big Bang, loop quantum gravity predicts a Big Bounce. As we go back in time, the Universe is contracting, density is increasing. However, density cannot increase forever. When the Universe is of the order of Planck size, quantum effects start operating and introduce a repulsive force, building up quantum degeneracy pressure. As the pressure builds up, a stage comes when the Universe bounces back.

It is important to realize that despite great promise, the theory of loop quantum gravity is far from complete. It is still developing. One problem is the time. As we now understand, space and time have no separate existence; they are intermingled and are part of the space-time continuum. In loop quantum gravity, while the space is quantized, the time is not. And what is even more worrying is that, at present, we do not even know how to quantize it. The other problem of loop quantum gravity is that as yet, it has no definite, testable prediction, not to speak of experimental verification. The granular structure of space will be revealed only at the Planck length scale of 10^{-33} cm. The length scale is beyond any present day accelerator or any accelerator conceivable in foreseeable future.

The second approach to quantum gravity is the string theory. Indeed, apart from quantizing gravity, string theory envisages to accomplish much more; it envisages unification of all the forces. For ages, scientists are trying to unify all the forces. In 1865, the first step was taken by James Clerk Maxwell when he unified electricity and magnetism. 1920 onwards, the unification of gravity and electromagnetism (strong and weak forces were yet to be discovered) was the cherished dream of Einstein. He spent his later life in the vain attempt to unify electromagnetism with gravity. The next progress towards unification came in 1968, when three physicists, Sheldon Lee Glashow, Abdus Salam and Steven Weinberg developed the "electroweak theory" unifying electromagnetic force with the weak force. In 1979, the three physicists were awarded the Nobel Prize *"for their contributions to the theory of the unified weak and electromagnetic interaction between elementary particles, including,* inter alia*, the prediction of the weak neutral current."* Later, attempts were made to unify strong and electroweak interaction. The theory, called Grand Unified Theory, however, was not a success. It predicted the decay of the proton, which was not observed experimentally. The string theory now promises to fructify the long cherished dream of the scientists, to have one theory for all the fundamental forces of nature, one equation which will explain our world. String theory promised to be the theory of everything.

The basic idea of string theory is simple. In the Standard Model fundamental particles have no dimension, they are point particles. String theory endows them with a dimension — dimension of the length — particles are no longer point particles, rather they are string-like. Undoubtedly, the length scale is very small, of the order of Planck length $\sim 10^{-33}$ cm. The strings can be open (with two free ends) or closed (like a loop). Like a rubber band, strings can expand and shrink. They expand when they gain energy, shrink when they lose energy. They can also vibrate. Now, a string of a guitar can vibrate in different modes, each mode producing different kinds of sounds. Similarly, fundamental strings can vibrate in different modes and these different modes of vibration produce different kinds of particles. In other words, in string theory, all the fundamental particles are nothing but different modes of vibration of our tiny strings of Planck length. Truly, a remarkable outcome. Strings are endowed with two properties: string tension or the energy per unit length of the string and string coupling constant. The coupling constant determines string interactions. An open string can fragment into two open strings; two open strings can join to form a single open string; a closed string (loop) can fragment into two loops etc. When the coupling constant is large, the strings interact strongly; when the coupling constant is small, the interaction is weak.

However, trouble started when one tries to apply quantum mechanics and relativity to the strings. Several catastrophes happen:

(i) In our ordinary 4-dimensional world (with three spatial dimensions and one temporal dimension), string theory is not consistent with quantum mechanics and relativity. The theory is consistent with quantum mechanics and relativity either in 10 dimensions or in 26 dimensions.

(ii) The theory will contain all types of fundamental particle species, only if it is supersymmetric. Supersymmetry is a symmetry where for every possible fermionic particle, there is a bosonic particle and *vice versa*. In other words, supersymmetry introduces a superpartner for each and every particle. By convention, superpartners of fermions begin with an "s" and those of bosons end with "ino." Thus string theory predicts that for the fermionic electron, there is a bosonic superpartner, "selectron." For the photon, which is a boson, string theory predicts a fermionic superpartner "photino," etc. However, in nature, we do not see the superpartners. There is an additional problem: supersymmetry can be incorporated in five different ways, giving five types of string theory.

These two problems would have killed the string theory if not for a gracing factor. Among the many vibrational modes of a supersymmetric string or in short, superstring, there is one massless spin-2 mode. In field theoretic model of gravity, matter particles interact by exchanging graviton — a spin-2 massless boson. The superstring theory automatically contains graviton, the mediator of gravitational interaction. Scientists were exhilarated; here is a theory, which has the potential for unifying all the forces, a theory of everything.

What to do with the catastrophes mentioned above? Well, the problem of supersymmetry can be circumvented by postulating rather high masses for the superpartners. Assume that supersymmetry is a symmetry of nature at very high energy e.g. at the early Universe. It is broken at lower energy, as in our present Universe. The superpartners would have been produced in laboratory experiments if we could collide particles with sufficient energy. As they are yet to be seen in experiments, presumably, we have not collided particles with enough energy. But what about the extra dimensions required by the string theory? The concept of extra dimensions was not new to scientists. In early 20s, Theodor Kaluza, a German mathematician and physicist, and Oscar Klein, a Swedish physicist, solved general relativity in five dimensions. The extra dimension can be compactified, i.e. curled up in a small circle to make it invisible to the world. The solution reproduces Einstein's gravity in ordinary 4 dimensional world, and additionally, the presence of the fifth dimension gave an equation describing electromagnetism. Thus, a theory of pure gravity in 5 dimensions is equivalent to a theory of gravity plus electromagnetism in 4 dimensions. Einstein, when learned about Kaluza's result was euphoric, but the euphoria was short lived. It turned out that the solution is unstable. Give a minimal disturbance and the circular extra dimension collapses to a singularity. String theorists now have revived Kaluza's idea to make the extra dimensions small. Fortunately for them, Eugenio Calabi, an Italian-American mathematician, and Shing-Tung Yau, a Chinese-American mathematician, discovered and extensively studied a six-dimensional space, which was named after them, the Calabi–Yau space. If the six extra dimensions of the 10-dimensional string theory are curled up on the six-dimensional Calabi–Yau space, the instability problem faced in the Kaluza–Klein model disappears. However, there is a problem. The Calabi–Yau space is not unique, there are many, may be hundred thousands of Calabi–Yau spaces, on each of which the extra dimensions can be curled up. Each of the five superstring theories can be curled up on hundred thousands of Calabi–Yau spaces, resulting in hundred thousands of string theories, each of which is different.

At this point string theorists discovered that the five versions of the superstring theory are solutions of an 11-dimensional theory called M-theory. What "M" stands for is uncertain, it could be "magic," "mystery" or "matrix," the exact meaning will be apparent when more complete theory is available. M can also stand for membrane, because, in one way or another, the theory contains surfaces or membranes. M-theory generalizes the notion of a point particle to higher dimensions by introducing the concept of p-brane, an object of p-dimensional spatial extent. Thus a 0-brane is a point particle, a 1-brane is a string, a 2-brane is a surface, a 3-brane is a volume etc. A special class of p-brane is called D-brane (D stands for a complex boundary condition called the Dirichlet condition, which I shall not elaborate). In M-theory, apart from the fundamental strings, we have fundamental objects called D-branes. A string in 11 dimensions can be viewed as a 2-brane in 9 dimensions, or a 3-brane in 8 dimensions etc. The introduction of branes enriches the M-theory. The five versions of the 10-dimensional string theory became related to one another; certain pairs of them are dual to each other. To understand the concept of duality, consider the following situation, say, Ted and Robin went to a jungle and had certain experiences. If asked to narrate their experiences, their narration will not be identical, but Ted's experiences will relate in one way or another with Robin's experiences. A clever man can deduce that they are narrating the same incidence. String theorists discovered two types of dualities: topological or T-duality and strong–weak or S-duality. T-duality exists between certain pairs of the five types of superstring theory: a string theory curled on a circle of radius R is dual to the string theory curled on a circle of radius $1/R$. S-duality is more subtle. By S-duality, a string theory with coupling constant g is dual to a string theory with coupling constant $1/g$. In other words, a strongly interacting string theory is equivalent to a weakly interacting string theory. How can two theories with drastically different coupling strengths behave identically? Scientists believe that it is an example of the little understood "emergent" phenomena. A large, complex system may exhibit properties unrelated to its constituents. For example, in the smallest level, everything is made up of a few fundamental particles like quarks, electrons etc., yet different things have very different properties. A living cell and a piece of iron, both are made of the same constituent matters — quarks, gluons etc., — yet their properties are completely different. In large, complex systems, a new property, unrelated to its constituents, can emerge.

In *Emperor's New Mind*, Roger Penrose classified the theories into three types: SUPERB, USEFUL and TENTATIVE. Euclidean geometry, Galilean dynamics, Maxwell's equations, Einstein's special and general relativity theories, quantum physics and quantum electrodynamics are all SUPERB, they have withstood the testing of time. The Standard Model of particle physics however comes under the category of USEFUL. It is a useful theory, but the formulation leaves much to be desired and also yet to be time tested. The string theory can at best be categorized as TENTATIVE. The formulation is not complete, far from experimental confirmation. However, even this tentative theory holds out several prospects for cosmology.

String theory gave us a clue to answer the question "What happened before the Big Bang?" As it is in loop quantum gravity, in string theory also, the initial Big Bang singularity is eliminated. The Universe's size is limited by the string scale, which, even though small, is not zero. We cannot collapse the Universe to infinite density. Since the initial singularity is avoided, physics can be traced back in time through the Big Bang into an earlier era, the pre-Big Bang era, where many of the initial conditions for the post-Big Bang are determined in a natural and dynamical way. Time reversal symmetry and certain other symmetries in string cosmology suggest that the present post-Big Bang cosmological phase must be preceded in time by an almost specularly symmetric phase occurring before the Big Bang. The scenario is depicted in Figure 6.3. The scenario is in line with the T-duality in string theory. According to T-duality, string theory compactified on a small circle is equivalent to a string theory compactified on a large circle. In some sense, large and small scales are, therefore, equivalent. Application of this idea to the Big Bang would suggest that if we trace the expansion far enough back in time, we are better off describing the Universe as becoming bigger again, rather than smaller. We may be living in a cyclic Universe, oscillating between a very-high-density Universe and very-low-density Universe. Very interestingly, the Universe is also cyclic in Hindu cosmology. How the Hindu sages arrived at this startling conclusion is uncertain, but in their vision, the world is created, destroyed, and recreated in an eternally repetitive series of cycles. There are many Gods in Hinduism, but the Holy triad — Brahma, Vishnu and Maheswara

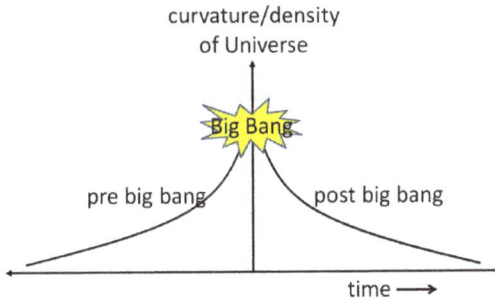

Figure 6.3. In string cosmology, standard post-Big Bang phase will be preceded by a dual phase of pre-Big Bang.

(or Shiva) — are considered to be supreme. According to Hindu mythology, Brahma creates the Universe, Vishnu nurtures it and once it is full with sin, Maheswara destroys it, again to be recreated by Brahma.

String theory also has the potential to illuminate on the dark matter and dark energy, two concepts poorly understood presently. We have seen evidence of dark matter, however we have no idea about the constituents of the dark matter. In the large list of particles in the string theory, there is one, generically named, lightest super-symmetric particle that is stable and neutral i.e. chargeless. It is an ideal candidate for the dark matter. The theory may also shed light on the dark energy. In string cosmology, as consequences of wrapping up the extra dimensions, a scalar field appears in the theory. The field is called dilaton. Though there is no specific prediction, the dilaton field is a potential candidate for the dark energy.

If the string theory is correct in the present version, it has an interesting consequence. We may be living in a "multiverse." There are many thousands of parallel Universes and our Universe is only one of them. Multiverse is natural in string theory. Remember that six of the extra dimensions are to be curled up on Calabi–Yau space. Since there are hundred thousands of Calabi–Yau space, there are hundred thousands of string theories, each with different initial conditions. All of them are potential Universes. There are arguments for and against multiverse. As mentioned earlier, our Universe appears to be fine-tuned to support life. Proponents of multiverse argue that the problem of fine tuning of nature will be eliminated in multiverse. Consider a situation where you believed to be alone in a jungle. Suddenly, far away, a shot is fired and you are hurt. If you believe that nobody is out there to murder you, you will be surprised that a randomly fired bullet hits you. Now say, you learn that the jungle is full of people. You will be less surprised that the bullet hits you. Indeed, in a jungle full of people, if a bullet is fired randomly, it could hit anyone and that anyone could be you. Canadian philosopher John Leslie, in his book *Universes*, cites the above example to argue for the multiverse. Indeed, if there is only one Universe, our Universe, then it is surprising that out of many possibilities, our Universe evolved into a state suitable for life. However, if there are hundred thousands of Universes, then it is less surprising that one of them evolved into a condition suitable for sustaining life. Detractors however argue that multiverse will not be amenable to experiment. If we try to exempt the multiverse from experimental verification, it could erode public confidence in science and ultimately damage the study of fundamental physics. The purely speculative concept of multiverse finds an echo in religious cosmology. For example, in Hindu cosmology or Islamic cosmology one accepts the plurality of Universe. One category of Hindu sacred texts is called *Purana*, meaning old or ancient. Written in easy flowing text, the post-Vedic Puranas were compiled by ancient Hindus to teach the essence of Hinduism to the general populace. There are 18 Puranas, each alluding to the virtues of a specific God. One of the Puranas is Bhagavata Purana, extolling the God Vishnu. In Bhagavata Purana, the concept of multiverse appears again and again.

String theory is far from complete and there are several criticisms against it. One of the severest critics of string theory was Richard Feynman. He thought that the theory was crazy and was in the wrong direction. When asked why he did not like the theory, he replied,[1]

> "I don't like that they are not calculating anything. I don't like that they don't check their ideas. I don't like that for anything that disagrees with an experiment, they cook up an explanation — a fix-up to say "Well, it still might be true." For example, the theory requires ten dimensions. Well, maybe there's a way of wrapping up six of the dimensions. Yes, that's possible mathematically, but why not seven? When they write their equation, the equation should decide how many of these things are wrapped up, not the desire to agree with experiment. In other words, there's no reason whatsoever in superstring theory that it isn't eight of the ten dimensions that get wrapped up and that the result is only two dimensions, which would be completely in disagreement with experience. So the fact that it might disagree with experience is very tenuous, it doesn't prove anything; it has to be excused most of the time. It doesn't look right."

The questions raised by Feynman are yet to be answered. Indeed, both quantum theories of gravity — loop quantum gravity and string theory (with its promise to be the theory of everything) — are far from complete, and yet to make any contact with experiment. Scientists now believe that both are true in parts, and an ultimate theory of everything will come from a combination of the two. Presently, we don't know how it will be done but surely as it happened many times before, mankind will find the way. When it is done, we shall know more about our origin, our purpose of existence and in a way, we shall know the mind of God.

---

[1] Excerpted from *Superstrings: A Theory of Everything*, ed. P.C.W. Davies and Julian Brown, Cambridge University Press, 1988.

About the Author

Asis Kumar Chaudhuri is a theoretical physicist and presently leading the Nuclear Theory Group at the Variable Energy Cyclotron Centre, Kolkata; a premier research institute in India. He received his initial training at the prestigious Training School, Bhabha Atomic Research Centre, Mumbai. He obtained his doctoral degree from the University of Calcutta. His primary research interest is high energy nuclear physics, in particular, physics of quark gluon plasma. He has a healthy interest in other areas of Physics, e.g. chaos, early universe, physics beyond the Standard Model etc. and occasionally contributed scientifically in those areas. He has authored approximately 100 research articles and an advanced textbook, *A Short Course on Relativistic Heavy Ion Collisions*, published by the Institute of Physics, London. Presently, He lives in Kolkata with his wife and son. He may be contacted at asiskumarchaudhuri@gmail.com.

Index